OPPORTUNITIES IN BIOTRANSFORMATIONS

Papers presented at the Conference on Opportunities in Biotransformations held at Churchill College, University of Cambridge, UK, 3–5 April 1990, organised by SCI in collaboration with the Biotransformation Club.

OPPORTUNITIES IN BIOTRANSFORMATIONS

Edited by

L.G. COPPING

DowElanco Ltd, Wantage, Oxon, UK

R.E. MARTIN

Laboratory of the Government Chemist, Teddington, Middlesex, UK

J.A. PICKETT

Rothamsted Experimental Station, Harpenden, Herts, UK

C. BUCKE

School of Biotechnology, Polytechnic of Central London, London, UK

A.W. BUNCH

Biological Laboratory, University of Kent at Canterbury, UK

SCI
FOR THE APPLICATION
OF CHEMISTRY AND
RELATED SCIENCES

Published for the
SOCIETY OF CHEMICAL INDUSTRY
by
ELSEVIER APPLIED SCIENCE
LONDON and NEW YORK

ELSEVIER SCIENCE PUBLISHERS LTD
Crown House, Linton Road, Barking, Essex IG11 8JU, England

Sole Distributor in the USA and Canada
ELSEVIER SCIENCE PUBLISHING CO., INC.
655 Avenue of the Americas, New York, NY 10010, USA

WITH 31 TABLES AND 41 ILLUSTRATIONS

© 1990 SOCIETY OF CHEMICAL INDUSTRY

British Library Cataloguing in Publication Data

Opportunities in biotransformations
1. Biotechnology
I. Copping, Leonard G.   II. SCI
660.6

ISBN 1-85166-517-X

Library of Congress CIP data applied for

No responsibility is assumed by the Publisher for any injury and/or damage to persons or property as a matter of products liability, negligence or otherwise, or from any use or operation of any methods, products, instructions or ideas contained in the material herein.

Special regulations for readers in the USA

This publication has been registered with the Copyright Clearance Center Inc. (CCC), Salem, Massachusetts. Information can be obtained from the CCC about conditions under which photocopies of parts of this publication may be made in the USA. All other copyright questions, including photocopying outside the USA, should be referred to the publisher.

Printed in Great Britain at the University Press, Cambridge

PREFACE

Opportunities for the use of biotransformation in the production of chemicals lie both in finding novel routes to compounds which are currently produced by traditional chemistry and in biocatalytic synthesis of completely new chemical entities which are inaccessible via existing chemistry.

Advantages from applying biotransformation technology can arise from improved cost, greater environmental acceptability and milder conditions leading to high purity products and improved impurity profiles. The major advantage, however, is the specificity of enzyme systems, particularly their ability to catalyse both regiospecific and stereospecific reactions.

Nevertheless, success will come from combining biotechnology with traditional chemistry rather than competing with it. In the long term, catalysis via natural systems or from systems derived from nature will bridge the gap between organic chemistry and fermentative production of natural products.

This conference represents the first venture of SCI into the area of biotransformation, and it is appropriate that it should be organised in collaboration with the Biotransformation Club. We all hope that it will be the first of a series of such meetings.

The organisers would like to thank all delegates for attendance, all contributors for the excellence of their papers, and all industrial organisations for their generous contributions to the bursary fund. Contributions were received from:

Beecham Pharmaceuticals
BP Research
Xenova Limited
Duphar BV

To all of these we extend our thanks.

L.G. COPPING
R.E. MARTIN
J.A. PICKETT
C. BUCKE
A.W. BUNCH

CONTENTS

viii

ix

LIST OF CONTRIBUTORS

M.W. Adlard, School of Biological and Health Sciences, Polytechnic of Central London, 115 New Cavendish Street, London W1M 8JS, UK (p. 47)

T.J. Bartlett, School of Biological and Health Sciences, Polytechnic of Central London, 115 New Cavendish Street, London W1M 8JS, UK (p. 47)

B. Berger, Institute of Organic Chemistry and Christian Doppler Laboratory for Chiral Compounds, Graz University of Technology, Stremayrgasse 16, A-8010 Graz, Austria (p. 100)

W.H.J. Boesten, DSM Research, Bio-organic Chemistry Section, PO Box 18, 6160 MD Geleen, The Netherlands (p. 148)

L. Brown, Departments of Bioscience & Biotechnology and Pure & Applied Chemistry, University of Strathclyde, Glasgow G1 1XW, UK (p. 81)

Q.B. Broxterman, DSM Research, Bio-organic Chemistry Section, PO Box 18, 6160 MD Geleen, The Netherlands (p. 148)

C. Bucke, School of Biological and Health Sciences, Polytechnic of Central London, 115 New Cavendish Street, London W1M 8JS, UK (p. 47)

A.W. Bunch, The Biological Laboratory, The University, Canterbury, Kent CT2 7NJ, UK (p. 88)

P.S.J. Cheetham, Unilever Research, Colworth Laboratory, Colworth House, Sharnbrook, Bedford MK44 1LQ, UK (p. 177)

B.C. Cunningham, Department of Biomolecular Chemistry, Genentech, Inc., 460 Pt San Bruno Blvd, South San Francisco, California 94080, USA (p. 32)

H. Dalton, Department of Biological Sciences, Warwick University, Coventry, West Midlands CV4 7AL, UK (pp. 72, 74)

I.C.M. Dea, Biotechnology Unit, Leatherhead Food Research Association, Randalls Road, Leatherhead, Surrey KT22 7RY, UK (p. 59)

J.A.M. de Bont, Agricultural University, Division of Industrial Microbiology, PO 8129, 6700 EV Wageningen, The Netherlands (p. 78)

A.M. Denholm, Department of Biological Sciences, University College of Wales, Aberystwyth, Dyfed SY23 3DA, UK (p. 195)

G.A. Dervakos, SERC Centre for Biochemical Engineering, Department of Chemical and Biochemical Engineering, University College London, Torrington Place, London WC1E 7JE, UK (p. 5)

N.M. Dixon, Department of Biological Sciences, University College of Wales, Aberystwyth, Dyfed SY23 3DA, UK (p. 195)

C. Evans, Enzymatix Ltd, Cambridge Science Park, Milton Road, Cambridge, UK (p. 105)

K. Faber, Institute of Organic Chemistry and Christian Doppler Laboratory for Chiral Compounds, Graz University of Technology, Stremayrgasse 16, A-8010 Graz, Austria (p. 100)

G.R. Fenwick, AFRC Institute of Food Research, Colney Lane, Norwich, Norfolk NR4 7UA, UK (p. 44)

G. Fülling, Hauptlaboratorium der Hoechst AG, 6230 Frankfurt/Main 80, Federal Republic of Germany (p. 186)

I. Gill, University of Reading, Reading, Berkshire RG2 9AT, UK (p. 40)

S.E. Godtfredsen, Novo Nordisk, DK-2880 Bagsvaerd, Denmark (p. 17)

H. Griengl, Institute of Organic Chemistry and Christian Doppler Laboratory for Chiral Compounds, Graz University of Technology, Stremayrgasse 16, A-8010 Graz, Austria (p. 100)

P.J. Halling, Departments of Bioscience & Biotechnology and Pure & Applied Chemistry, University of Strathclyde, Glasgow G1 1XW, UK (p. 81)

A.B. Hanley, AFRC Institute of Food Research, Colney Lane, Norwich, Norfolk NR4 7UA, UK (p. 44)

R.J. Hart, Biotechnology Unit, Leatherhead Food Research Association, Randalls Road, Leatherhead, Surrey KT22 7RY, UK (p. 59)

S. Hartmans, Agricultural University, Division of Industrial Microbiology, PO 8129, 6700 EV Wageningen, The Netherlands (p. 78)

M. Helwig, Lonza AG, CH-3930 Visp, Switzerland (p. 67)

F.W.J.M.M. Hoeks, Lonza AG, CH-3930 Visp, Switzerland (p. 67)

E.W. Holla, Hauptlaboratorium der Hoechst AG, 6230 Frankfurt/Main 80, Federal Republic of Germany

J.R. Hunt, Department of Biological Sciences, Warwick University, Coventry, West Midlands CV4 7AL, UK (p. 72)

E.W. James, Department of Biological Sciences, University College of Wales, Aberystwyth, Dyfed SY23 3DA, UK (p. 195)

G.A. Johnston, Departments of Bioscience & Biotechnology and Pure & Applied Chemistry, University of Strathclyde, Glasgow G1 1XW, UK (p. 81)

S.E. Jones, Department of Biological Sciences, Warwick University, Coventry, West Midlands CV4 7AL, UK (p. 74)

L. Jørgensen, University of Reading, Reading, Berkshire RG2 9AT, UK (p. 40)

J. Kamphuis, DSM Research, Bio-organic Chemistry Section, PO Box 18, 6160 MD Geleen, The Netherlands (p. 148)

A. Kasal, Institute of Organic Chemistry and Biochemistry, Czechoslovak Academy of Sciences, 166 10 Prague, Czechoslovakia (p. 50)

D.B. Kell, Department of Biological Sciences, University College of Wales, Aberystwyth, Dyfed SY23 3DA, UK (p. 195)

R. Keller, Hauptlaboratorium der Hoechst AG, 6230 Frankfurt/Main, Federal Republic of Germany (p. 186)

M. Kloosterman, DSM Research, Bio-organic Chemistry Section, PO Box 18, 6160 MD Geleen, The Netherlands (p. 148)

K. Königsberger, Institute of Organic Chemistry and Christian Doppler Laboratory for Chiral Compounds, Graz University of Technology, Stremayrgasse 16, A-8010 Graz, Austria (p. 100)

I.A. Kozlov, University of Reading, Reading, Berkshire RG2 9AT, UK (p. 40)

B.A. Law, AFRC Institute of Food Research, Shinfield, Reading RG2 9AT, UK (p. 40)

C. Lee, Enzymatix Ltd, Cambridge Science Park, Milton Road, Cambridge, UK (p. 105)

P. Lehky, Lonza AG, CH-3930 Visp, Switzerland (p. 67)

J.A. Lewis, AFRC Institute of Food Research, Colney Lane, Norwich, Norfolk NR4 7UA, UK (p. 44)

M.D. Lilly, SERC Centre for Biochemical Engineering, Department of Chemical and Biochemical Engineering, University College London, Torrington Place, London WC1E 7JE, UK (p. 5)

E.M. Meijer, DSM Research, Bio-organic Chemistry Section, PO Box 18, 6160 MD Geleen, The Netherlands (p. 148)

H.-P. Meyer, Lonza AG, CH-3930 Visp, Switzerland (p. 67)

J.G. Morris, Department of Biological Sciences, University College of Wales, Aberystwyth, Dyfed SY23 3DA, UK (p. 195)

K. Mosbach, Pure and Applied Biochemistry, Chemical Center, University of Lund, PO Box 124, S-221 00 Lund, Sweden (p. 1)

H. Murakami, Biotechnology Laboratory, Takarazuka Research Center, Sumitomo Chemical Co. Ltd, Takarazuka, Hyogo 665, Japan (p. 23)

K.G.I. Nilsson, Carbohydrates International AB, Arlöv, Sweden. Present address: Chemical Center, University of Lund, PO Box 124, S-221 00 Lund, Sweden (p. 131)

H. Ohkawa, Biotechnology Laboratory, Takarazuka Research Center, Sumitomo Chemical Co. Ltd, Takarazuka, Hyogo 665, Japan (p. 23)

K.R. Parsley, AFRC Institute of Food Research, Colney Lane, Norwich, Norfolk NR4 7UA, UK (p. 44)

J. Pokorná, Institute of Organic Chemistry and Biochemistry, Czechoslovak Academy of Sciences, 166 10 Prague, Czechoslovakia (p. 50)

D. Quarroz, Lonza AG, CH-3930 Visp, Switzerland (p. 67)

R.A. Rastall, School of Biological and Health Sciences, Polytechnic of Central London, 115 New Cavendish Street, London W1M 8JS, UK (p. 47)

D.W. Ribbons, Department of Biotechnology, Imperial College of Science, Technology and Medicine, London SW7 2AY, UK (p. 119)

S.M. Roberts, Department of Chemistry, University of Exeter, Stocker Road, Exeter, Devon EX4 4QD, UK (pp. 105, 140)

S.D. Roller, Biotechnology Unit, Leatherhead Food Research Association, Randalls Road, Leatherhead, Surrey KT22 7RY, UK (p. 59)

T. Sakaki, Biotechnology Laboratory, Takarazuka Research Center, Sumitomo Chemical Co. Ltd, Takarazuka, Hyogo 665, Japan (p. 23)

D.B. Sarney, AFRC Institute of Food Research, Shinfield, Reading RG2 9AT, UK (p. 40)

H.E. Schoemaker, DSM Research, Bio-organic Chemistry Section, PO Box 18, 6160 MD Geleen, The Netherlands (p. 148)

M.T. Scott, Agricultural Products Department, Experimental Station, E.I. du Pont de Nemours & Company, Wilmington, Delaware 19880-0402, USA (p. 95)

M. Shibata, Biotechnology Laboratory, Takarazuka Research Center, Sumitomo Chemical Co. Ltd, Takarazuka, Hyogo 665, Japan (p. 23)

N.F. Shipston, The Biological Laboratory, The University, Canterbury, Kent CT2 7NJ, UK (p. 88)

K.C. Srivastava, Michigan Biotechnology Institute, PO Box 27609, Lansing, Michigan 48909, USA (p. 53)

B. Stieglitz, Agricultural Products Department, Experimental Station, E.I. du Pont de Nemours & Company, Wilmington, Delaware 19880-0402, USA (p. 95)

C.J. Suckling, Department of Pure and Applied Chemistry, University of Strathclyde, 295 Cathedral Street, Glasgow G1 1XL, UK (pp. 36, 81)

A. Sutherland, Department of Chemistry, University of Exeter, Stocker Road, Exeter, Devon EX4 4QD, UK (p. 105)

S.J. Swinton, Biotechnology Unit, Leatherhead Food Research Association, Randalls Road, Leatherhead, Surrey KT22 7RY, UK (p. 59)

S. Taylor, Enzymatix Ltd, Cambridge Science Park, Milton Road, Cambridge, UK (p. 105)

S.C. Taylor, ICI Biological Products, PO Box 1, Billingham, Cleveland TS23 1LB, UK (p. 170)

S. Thomas, Enzymatix Ltd, Cambridge Science Park, Milton Road, Cambridge, UK (p. 105)

R.H. Valivety, Departments of Bioscience & Biotechnology and Pure & Applied Chemistry, University of Strathclyde, Glasgow G1 1XW, UK (p. 81)

M.J. van der Werf, Agricultural University, Division of Industrial Microbiology, PO 8129, 6700 EV Wageningen, The Netherlands (p. 78)

E.N. Vulfson, AFRC Institute of Food Research, Shinfield, Reading RG2 9AT, UK (p. 40)

J.A. Wells, Department of Biomolecular Chemistry, Genentech, Inc., 460 Pt San Bruno Blvd, South San Francisco, California 94080, USA (p. 32)

D.A. Widdowson, Department of Chemistry, Imperial College of Science, Technology and Medicine, London SW7 2AY, UK (p. 119)

R. Wisdom, Enzymatix Ltd, Cambridge Science Park, Milton Road, Cambridge, UK (p. 105)

J.M. Woodley, SERC Centre for Biochemical Engineering, Department of Chemical and Biochemical Engineering, University College London, Torrington Place, London WC1E 7JE, UK (pp. 5, 63)

L.F.J. Woods, Biotechnology Unit, Leatherhead Food Research Association, Randalls Road, Leatherhead, Surrey KT22 7RY, UK (p. 59)

Y. Yabusaki, Biotechnology Laboratory, Takarazuka Research Center, Sumitomo Chemical Co. Ltd, Takarazuka, Hyogo 665, Japan (p. 23)

J. Zajíček, Institute of Organic Chemistry and Biochemistry, Czechoslovak Academy of Sciences, 166 10 Prague, Czechoslovakia (p. 50)

NOVEL APPROACHES OF POTENTIAL USE IN BIOTRANSFORMATION

KLAUS MOSBACH
Pure and Applied Biochemistry
Chemical Center, University of Lund
P.O. Box 124, S-221 00 Lund, SWEDEN

In this brief article I do not intend to cover in an overview fashion the large number of new biotransformation reactions that have emerged during the last few years, whether on a laboratory or pilot plant scale or those that already are used as processes. It suffices to refer to some recent articles, symposia proceedings, or books published in this exciting area (1-5). Rather, I would like to address aspects relating to novel approaches that are potentially useful in this area, and particularly stress work carried out in our laboratory.

For instance, we have all witnessed the increasing interest in the use of enzymes in organic solvents and the number of papers published is increasing rapidly (2, 3, 6-9). It has likewise been shown that biosensors can be usefully applied to monitor enzymic reactions carried out in organic solvents (10). Related in a sense to these approaches are reports in the literature describing reactions carried out in organic solvents with enzymes or proteins, with altered specificity. Here we would like to refer to recent works by Klibanov (11) and our group; in the latter case a kind of bio-imprinting approach was used, allowing α-chymotrypsin to accept a D-amino acid to form the corresponding D-tryptophan ethyl ester (12).

The increased use of thermostable enzymes to carry out bio-transforming reactions is also noteworthy. Another aspect of great interest is that of coenzyme recycling, as a large number of enzymes require coenzymes, in particular the adenine coenzymes ATP, NAD, and NADP (13-15).

A major new development that has aroused great interest is the area of catalytic antibody technology (16, 17). To what degree ribozymes might be practically useful remains to be seen, although the concept as such is exciting (18).

Obviously, the technique of protein engineering will have an ever-increasing impact on the general area of biotransformation. The rational de novo design of enzymes represents a great potential. (See the contribution by the group of Genentech in this symposium). In this context we would like to refer to attempts made using recombinant DNA-technology to prepare peptides made up of repetitive catalytic units expressing some esterolytic activity (19).

Most of the biotransformational work carried out until the present has primarily dealt with compounds such as amino acids, esters, steroids or alkaloids. A group of compounds of potential medicinal interest is the biologically active carbohydrates (20, 21).

Remarkable work on large scale enzymic production of acrylamide using the enzyme nitrilase present in cells with production capacities of thousands of tons per year was recently reported (22).

It is to be expected that enzymes present in plant cells and useful in biotransformation of substances such as alkaloids or steroids will be made to be overproduced, in addition to making plants produce hormones, new enzymes, or antibodies (23). Likewise, interesting enzymes normally present in plants will be transferred to microorganisms.

An area that has attracted a lot of attention in our institute in recent years is that of designing bi- and polyfunctional enzymes by gene fusion. As a number of biotransformation processes include two or more enzymic steps, such as starch degradation or the formation of 7-amino cephalosporanic acid from cephalosporin C involving oxidase and acylase (24), such systems could be usefully applied either in vitro or in vivo. Artificial enzyme hybrids obtained by gene fusion in our laboratory include β-galactosidase - galactose kinase, β-galactosidase - galactose dehydrogenase as well as β-galactosidase - galactose kinase - galactose dehydrogenase (25, 26).

Using the same kind of approach hormones or enzymes are increasingly being modified to carry affinity tags or tails, facilitating subsequent bioseparation (27, 28).

Enzymes are increasingly being used for production of optically pure compounds. An alternative approach somewhat inspired by the "lock and key" concept of enzymes is the molecular imprinting technology; here, the goal is to prepare tailor-made separation material with particular emphasis on distinguishing between optical antipodes. Although it still is in its infancy, the technique shows great potential. Using this approach, we and others can effect efficient separation of racemic mixtures of some groups of compounds (29). At a later stage maybe artificial catalysts can be obtained using this approach (30).

In closing this introduction, in which I have restricted myself mainly to our own work and for obvious reasons could not cover the enormous and outstanding work by the many other groups working in the area, I would like to express the conviction that many of these novel approaches towards improved enzymes/cells or artificial biocatalysts will have a great impact, leading to the increased use of biotransformation processes.

REFERENCES

1. Mosbach, K. Ed. Meth. Enzymol. vols. 135, 136, 1987.

2. Klibanov, A.M. Enzymatic catalysis in anhydrous organic solvents. Trends Biochem. Sci., 1989, 14, 141-144.

3. Klibanov, A.M. Enzymes that work in organic solvents. Chemtech., 1986, 16, 354-359.

4. Laskin, A.I., Mosbach, K., Thomas, D. and Wingard Jr., L.B. Eds. Enzyme Engineering 8, 1987, Ann. New York Acad. Sci., vol. 501.

5. Blanch, H.W. and Klibanov, A.M. Eds. Enzyme Engineering 9, 1988, Ann. New York Acad. Sci., vol. 542.

6. Lilly, M.D., Harbron, S. and Narendranathan, T.J., Two-liquid phase biocatalytic reactors. Meth. Enzymol., 1987, 136, 138-149 and Lilly, M.D., this volume.

7. Fukui, S. and Tanaka, A., Optical resolution of dl-menthol by entrapped biocatalysts. Meth. Enzymol., 1987, 136, 293-302.

8. Laane, C., Tramper, J. and Lilly, M.D. Eds. Biocatalysis in organic media. 1987, Elsevier, Amsterdam (and previous volumes in the same series).

9. Mozhaev, V.V., Khmelnitsky, Y.L., Sergeeva, M.V., Belova, A.B., Klyachko, N.L., Levashov, A.V. and Martinek, K., Catalytic activity and denaturation of enzymes in water organic cosolvent mixtures - chymotrypsin and laccase in mixed water alcohol, water glycol and water formamide solvents. Eur. J. Biochem., 1989, 184, 597-602.

10. Stasinská, B., Danielsson, B. and Mosbach, K., The use of biosensors in bioorganic synthesis: Peptide synthesis by immobilized α-chymotrypsin assessed with an enzyme thermistor. Biotechnol. Techniques, 1989, 3, 281-288.

11. Braco, L., Dabulis, K. and Klibanov, A.M., Production of abiotic receptors by molecular imprinting of proteins. Proc. Natl. Acad. Sci., 1990, 87, 274-277.

12. Ståhl, M., Månsson, M.-O. and Mosbach, K., The synthesis of a D-amino acid ester in organic media with chymotrypsin modified by a bio-imprinting procedure. Biotechnol. Lett., March 1990, in press.

13. Månsson, M.-O. and Mosbach, K., Immobilized pyridine nucleotide coenzymes. In Coenzymes and Cofactors, Eds. D. Dolphin, P. Poulsen and O. Avramovic, Wiley-Interscience, New York, 1987, pp. 217-273.

14. Lowe, C.R., Immobilized coenzymes., Trends Biol. Sci., 1978, 3, 134-137 and in this volume.

15. Vasic-Racki, Dj., Jonas, M., Wandrey, C., Hummel, W. and Kula, M.R., Continuous (R)-mandelic acid production in an enzyme membrane reactor. Appl. Microbiol. Biotechnol., 1989, 31, 215-222.

16. Lerner, R.A. and Tramontano, A., Antibodies as enzymes. Trends Biochem. Sci., 1987, 12, 427-430.

17. Schultz, P.G., Catalytic antibodies. Acc. Chem. Res., 1989, 22, 287-294.

18. Zaug, A.J. and Cech, T.R., The intervening sequence RNA of *Tetrahymena* is an enzyme. Science, 1986, 231, 431-435.

19. Bülow, L. and Mosbach, K., The expression in *E. coli* of a polymeric gene coding for an esterase mimic catalyzing the hydrolysis of *p*-nitrophenyl esters. FEBS Lett., 1987, 210, 147-152.

20. Nilsson, K.G.I., Enzymatic synthesis of oligosaccharides. Trends Biotechnol., 1988, 6, 256-264 and in this volume.

21. Hedbys, L., Johansson, E., Mosbach, K., Larsson, P.-O., Gunnarsson, A., Svensson, S. and Lönn, H., Synthesis of Galβ1-3GlcNAcβ-SEt by an enzymatic method comprising the sequential use of β-galactosidases from bovine testes and *Escherichia coli*. Clycoconjugate J., 1989, 6, 161-168.

22. Nagasawa, T. and Yamada, H., Microbial transformations of nitriles. Tibtech., 1989, 73, 153-158

23. Hiatt, A., Cafferkey, R. and Bowdish, K., Production of antibodies in transgenic plants. Nature, 1989, 342, 76-78.

24. Szwajcer, E. and Mosbach, K., Isolation and partial characterization of a D-amino acid oxidase active against cephalosporin C from the yeast *Trigonopsis variabilis*. Biotechnol. Lett., 1985, 7, 1-7.

25. Bülow, L., Ljungcrantz, P. and Mosbach, K., Preparation of a soluble bifunctional enzyme by gene fusion. Bio/Technology, 1985, 3, 821-823.

26. Ljungcrantz, P., Carlsson, H., Månsson, M.-O., Buckel, P., Mosbach, K. and Bülow, L., Construction of an artificial bifunctional enzyme, β-galactosidase/galactose dehydrogenase, exhibiting efficient galactose channeling. Biochemistry, 1989, 28, 8786-8792.

27. Ljungquist, C., Breitholtz, A., Brink-Nilsson, H., Moks, T., Uhlén, M. and Nilsson, B., Immobilization and affinity purification of recombinant proteins using histidine peptide fusions. Eur. J. Biochem., 1989, 186, 563-569.

28. Persson, M., Bergstrand G:son, M., Bülow, L. and Mosbach, K., Enzyme purification by genetically attached polycysteine and polyphenyl-alanine affinity tails. Anal. Biochem., 1988, 172, 330-337.

29. Ekberg, B. and Mosbach, K., Molecular imprinting: A technique for producing specific separation materials. Trends Biotechnol., 1989, 7, 92-96.

30. Robinson, D.K. and Mosbach, K., Molecular imprinting of a transition state analogue leads to a polymer exhibiting esterolytic activity. J. Chem. Soc., Chem. Commun., 1989, 14, 969-970.

TWO−LIQUID PHASE BIOCATALYSIS: CHOICE OF PHASE RATIO

M. D. LILLY, G. A. DERVAKOS and J. M. WOODLEY
SERC Centre for Biochemical Engineering,
Department of Chemical and Biochemical Engineering,
University College London,
Torrington Place, London WC1E 7JE, UK.

ABSTRACT

Many enzymic and microbial reactions involving poorly water−soluble reactants or products require both an aqueous and an organic phase to be present. The ratio of the volumes of the two liquid phases has a marked effect on the mass transfer of reactants and products between the two phases, on the activity and stability of the catalysts and the subsequent recovery of catalyst and product(s). Some examples of such effects and guidelines for the choice of phase ratio are given.

INTRODUCTION

Many organic compounds of interest to the food, pharmaceutical and fine chemical industries have low solubilities in aqueous media. These may be water−immiscible liquids or alternatively solids which are soluble in organic solvents. Biotransformations of these compounds are complex because of the insoluble nature of the reactants and/or products. It is possible to operate reactors where the solid is in the form of an aqueous suspension [1,2] or the solubility of the solid has been increased by use of a water−miscible organic solvent [3,4]. The alternative is to carry out the reaction with an organic phase present in the reactor, consisting initially of either a water−immiscible liquid reactant or a water−immiscible solvent in which the reactant has been dissolved. Previously we have proposed a classification for such reactions [5]. Five main types of reaction system were identified based on the nature of the aqueous phase and the catalyst form. If the nature of the organic phase (as outlined above) is also included, then ten main types can be identified. We also showed that, excluding reactions where all the reaction components (reactant(s) and product(s)) are in the aqueous phase, twenty one different

$$C_aP/C_o \;=\; 1 \;-\; RP(1/k_oP \;+\; 1/k_a)/AC_o \qquad\qquad (1)$$

where C_o, C_a = concentrations of reactant in the bulk organic and
aqueous phases, respectively

P = reactant partition coefficient (organic/aqueous)

A = liquid/liquid interfacial area per unit aqueous volume

k_o, k_a = reactant mass transfer coefficients for films on
organic and aqueous sides of liquid/liquid interface

R = reaction rate per unit aqueous volume

i. Solvent

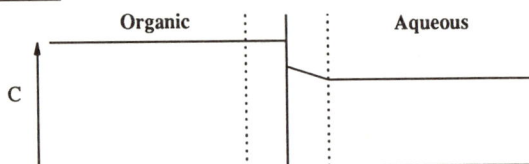

ii. Solute

Figure 1. Concentration profiles for a poorly water−soluble reactant with a water−soluble catalyst where the reactant is (i) the organic solvent and (ii) a solute dissolved in the organic phase.

The right−hand side of equation (1) is a measure of the effectiveness of the overall mass transfer and ranges between practically zero (high mass transfer resistance) and unity (negligible mass transfer resistance, i.e. kinetically controlled). Furthermore, A, k_a and k_o are all functions of phase ratio, agitation conditions, reactor geometry and physical properties of the two phases, especially density, viscosity and interfacial tension. In a recent paper we have shown that such an analysis is valid for these systems [9]. Hence, the choice of phase ratio is crucial to reactor design and operation.

distributions of reactant(s) and product(s) were possible for reactions involving up to two reactants and two products. Of these, four represent the situation where all the reaction components are predominantly present in the organic phase and a discrete aqueous phase may not be present. Examples of such reactions are fat interesterification [6,7] and a range of transesterifications [8]. Here, a small amount of water is essential in order to maintain catalytic activity but too much leads to a dominance of hydrolysis rather than the synthetic esterification. Control of the water content presents a challenging problem. The remaining seventeen component distributions are reactions where both aqueous and organic phases must be present for one or more of the following reasons:− (a) one or more of the reactants and products partition predominantly into the aqueous phase, (b) water is one of the reactants, and (c) the biocatalyst requires a discrete aqueous phase to function.

The presence of two liquid phases results in an interface between them across which transfer of reactant(s) and product(s) may occur and which will therefore have an important influence on the behaviour of such two−liquid phase reactors. In this paper we discuss the factors which influence the choice of phase ratio and its effect on the operation and performance of biocatalytic reactors. We have defined the phase ratio (Φ) as the fraction of the total reaction volume occupied by the organic phase.

THEORETICAL

The concentration profiles of a poorly water−soluble reactant at steady state in a two−liquid phase system where the organic phase initially consists of reactant, or of a solvent in which reactant is dissolved, are shown in Figures 1.i and 1.ii respectively. The discontinuities in the profiles at the liquid/liquid interface represent the partitioning of reactant. These profiles assume that both the organic and aqueous phases are well mixed and that the catalyst is dissolved in the aqueous phase. We have found such assumptions to hold in studies of enzymatic biphasic catalysis [9]. Where the catalyst is in a solid form, i.e. a cell or immobilised enzyme or cell, then there will be further resistances to mass transfer from the aqueous phase to the surface of the catalyst and into the catalyst if it is a porous immobilised catalyst. An equation has been derived [10] to describe mass transfer in such systems and may be simplified for the case where the organic phase initially consists of a water−immiscible solvent in which the reactant is dissolved and the catalyst is an enzyme dissolved in the aqueous phase. This case is shown in Figure 1.ii and may be described by the following equation:−

ENGINEERING CONSIDERATIONS

Operational constraints

Certain characteristics of the biotransformation may reduce the freedom to choose the phase ratio used. For instance, if the reaction results in the net production (e.g. racemic ester hydrolysis) or removal of hydrogen ions then it becomes necessary to control the reaction pH at the optimal value. When the initial amount of reactant is small relative to the aqueous volume then a buffer can be used if this is not too expensive and does not interfere with the reaction or subsequent product recovery. Often, it will be necessary to add acid or base in response to changes in pH measured by an electrode. However, it is difficult to make such measurements if the organic phase is the continuous phase so that an aqueous continuous phase would normally be used.

Any additions of acidic or alkaline solutions or other aqueous liquids will change the phase ratio and should be kept to the minimum. However, the rapid dispersion of high concentrations of acid or alkali added to the reactor is essential to avoid local regions of low or high pH causing catalyst inactivation and is most easily achieved when the aqueous phase is the continuous phase.

In the Theoretical section above we considered only reactant transfer. When a product is inhibitory to the catalyst and its partition coefficient (P) is greater than unity, then removal of the product from the aqueous phase becomes important. Not only the overall product mass transfer coefficient but also the phase ratio, together with the partition coefficient, are crucial in determining the aqueous phase product concentration and hence the reaction rate [11]. Thus, for a poorly water—soluble product a reduction in inhibition is generally favoured by a high phase ratio.

Mass transfer

The rate of mass transfer of a reactant between the liquid phases is a function of the interfacial area per unit volume of aqueous phase, A, and the organic and aqueous film mass transfer coefficients, k_o and k_a, in Equation 1. In an analogous manner, the rate of product transfer is a function of the interfacial area per unit volume of organic phase as well as the film mass transfer coefficients.

With an aqueous continuous phase, the interfacial area rises with the phase ratio for given agitation conditions. However, a point is reached at which phase inversion occurs so that the organic phase becomes the continuous phase. The phase ratio at which this transition occurs depends on the properties of the two liquid phases and the nature of the catalyst. For instance, high aqueous concentrations of microorganisms may increase the phase ratio at which inversion takes place. As the phase ratio is increased further beyond the inversion point, the interfacial area falls, reducing the rate of mass transfer. Maximum interfacial area may be reached at a phase ratio below that for phase inversion, which occurs normally in the range, 0.4 — 0.7. It is advantageous, therefore, to operate the reaction sufficiently below the point of phase inversion to maximise mass transfer and avoid uncontrolled phase inversion. In highly agitated reactors where the aqueous phase is dispersed as droplets, the internal film mass transfer coefficient may be reduced because of poor mixing within small droplets [12]. In reactors with high aqueous phase biocatalyst concentrations, this may result in a decrease in the observed reaction rate [13].

It is important to emphasize that many other factors, including agitation rate, influence the interfacial area at a particular phase ratio. For example, the interfacial area is reduced when the interfacial tension and phase density differences are high.

The above discussion is concerned with reactions occurring in the bulk aqueous phase. In a previous paper [9] we have identified important differences between such reactions and those which occur at the liquid/liquid interface. In this latter class of reactions, the rate of reaction is a direct function of total interfacial area and hence the phase ratio.

In general good mass transfer is not so difficult to achieve in liquid/liquid mixtures as in gas/liquid systems due to the low interfacial tensions and low density differences. Therefore, we believe that scale—up of reactors will not be limited by mass transfer problems [14]. However, the required power input may be a function of phase ratio on account of changing fluid characteristics.

Product recovery

The phase ratio affects both the ease of phase separation and the recovery of product(s). At extreme values of phase ratio ($\Phi < 0.1$; $\Phi > 0.9$) phase separation is poor. At intermediate values of phase ratio better recovery of the continuous phase, whether this be aqueous or organic, is achieved. The improved recovery is counterbalanced by the lower product concentration obtained as the volume of the phase in which it is dissolved increases.

For products with a partition coefficient which is high ($P > 20$) or low ($P < 0.05$) it is only worthwhile recovering product from the organic or aqueous phase respectively. However, for products with intermediate values of partition coefficient, it may be necessary to recover from both liquid phases. If this is not done, then the amount of product lost in the unrecovered phase can be minimised by choosing a phase ratio in favour of the recovered phase. In selecting a phase ratio it is important to consider both the amount of product recovered and its concentration in the phase which will be processed further. This trade—off between amount and concentration is further complicated by the properties of the liquid phases.

These effects of phase ratio on the various engineering aspects of a biotransformation are summarised qualitatively in Figure 2. However, it is impossible to divorce these engineering considerations from those relating to the properties of the biological catalyst.

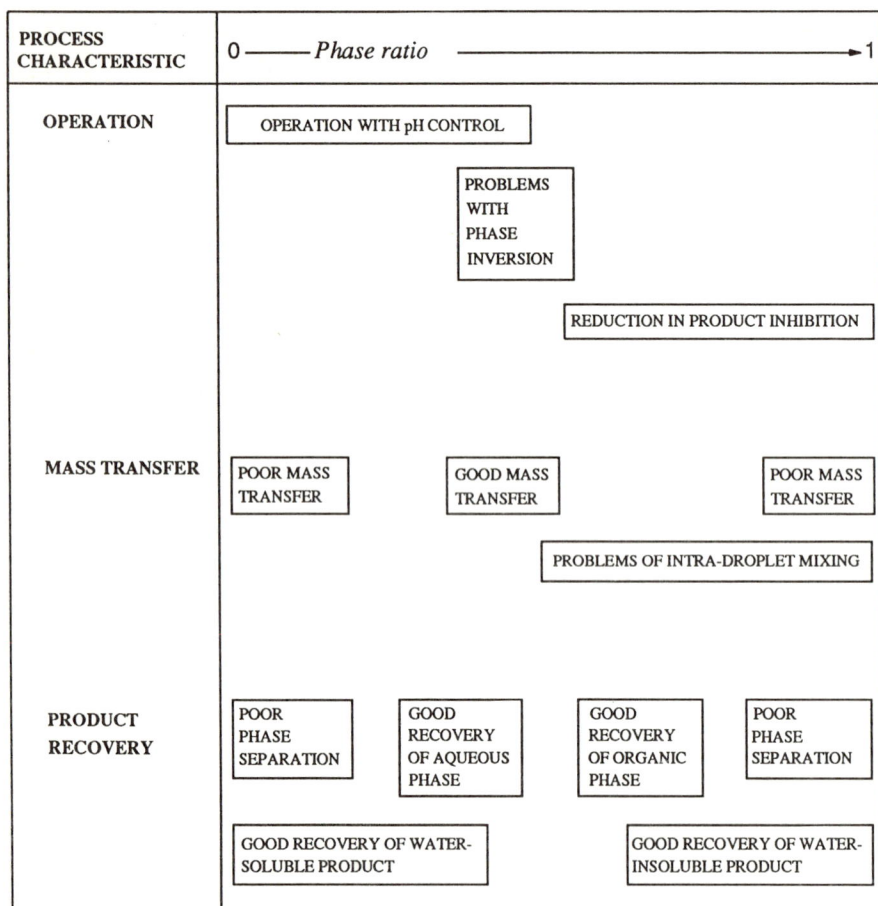

PROCESS CHARACTERISTIC	0 ——— *Phase ratio* ————————————————————→ 1			
OPERATION	OPERATION WITH pH CONTROL			
		PROBLEMS WITH PHASE INVERSION		
				REDUCTION IN PRODUCT INHIBITION
MASS TRANSFER	POOR MASS TRANSFER	GOOD MASS TRANSFER		POOR MASS TRANSFER
			PROBLEMS OF INTRA-DROPLET MIXING	
PRODUCT RECOVERY	POOR PHASE SEPARATION	GOOD RECOVERY OF AQUEOUS PHASE	GOOD RECOVERY OF ORGANIC PHASE	POOR PHASE SEPARATION
	GOOD RECOVERY OF WATER-SOLUBLE PRODUCT		GOOD RECOVERY OF WATER-INSOLUBLE PRODUCT	

Figure 2. Diagrammatic representation of the influence of phase ratio on operation, mass transfer and product recovery.

BIOLOGICAL CONSIDERATIONS

There is now considerable evidence that the organic phase can have a harmful effect on the stability of biological catalysts [15]. Although there appears to be a reasonable correlation for both enzymes and microorganisms between product formed over a given period and the log P (logarithm of the partition coefficient of the organic solvent in a standard octanol−water two−phase system) of the organic solvent used, there are marked differences when examined more closely. For instance, the characteristics of menthyl acetate hydrolysis by *Bacillus subtilis* and pig liver esterase differ in several respects [16]. At a constant agitation rate, the product formed in an hour by *B.subtilis* was optimal at a phase ratio of about 0.5 and fell at higher and lower values. In contrast, the enzymically−catalysed reaction showed a constant activity at low phase ratio but fell markedly at phase ratios above 0.4. Inactivation of the enzyme at the liquid/liquid interface can occur and some protection can be given by addition of an inert protein to the aqueous phase [17].

For biotransformations involving oxidation by microorganisms both the oxidising enzyme and a linked oxidation/reduction system must be active. For the Δ^1−dehydrogenation of hydrocortisone by *Arthrobacter simplex* the latter system is much more susceptible than the dehydrogenase itself to damage by solvents [18]. Much of this susceptibility results from contact between the bacteria and the liquid/liquid interface and a much more stable catalyst is obtained when the bacteria are immobilised within a porous structure, thereby protecting them from the interface [19].

Immobilisation, therefore, is a useful technique for improving the stability of biocatalysts, although the increase must justify the additional cost and immobilisation introduces further possible mass transfer problems as mentioned above as well as implications for process design. If the catalyst can be protected sufficiently then the freedom to choose the phase ratio on the basis of engineering criteria is greater.

In some cases reactants are highly toxic to microorganisms at concentrations below aqueous phase saturation. Toluene inhibits its own oxidation to the corresponding 1,2−dihydrodiol by *Pseudomonas putida* above about 50% aqueous saturation concentration [20]. Therefore, it is essential to supply toluene to the aqueous phase at such a rate that its concentration never exceeds that value. As oxygen is a second reactant it is possible to supply the reactant in the gas stream to the reactor but inevitably a large proportion of the toluene is lost in the outflowing gas [21]. A preferable method is to feed toluene directly into the aqueous suspension of bacteria with a high degree of agitation to ensure rapid dispersion and solution of the toluene [22].

The above examples emphasize the need to understand the effects on catalyst activity and stability of the organic phase, both that dissolved in the aqueous phase and the liquid/liquid interface, as such biological constraints may dominate the choice of phase ratio.

PRACTICAL EXAMPLES

Although much work has now been published on two—liquid phase biocatalysis, there are relatively few studies where the phase ratio has been varied and in some of these shaken tubes have been used rather than reactors for experimental work. Nevertheless, experimental studies have been done with enzymic and microbial reactions with and without added organic solvent (TABLE 1). In each case, except cholesterol oxidation, the range of phase ratios examined was sufficient for phase inversion to occur.

Despite the shortage of fundamental information on the behaviour of two—liquid phase reactors, at least ten processes have been scaled—up to pilot— or large—scale on an empirical basis and the reported values of phase ratio for some of these industrial applications is given in TABLE 2.

TABLE 1

Experimental studies on the effect of phase ratio upon biocatalytic activity

Reaction	Biocatalyst	ϕ	Ref
Without added solvent			
menthyl acetate hydrolysis	*Bacillus subtilis*	0.05 - 0.7	[23]
		0.1 - 0.9	[16]
	pig liver esterase	0.2 - 0.9	[16]
octadiene epoxidation	*Pseudomonas putida*	0.1 - 0.6	[24]
With added solvent			
cholesterol oxidation (in tetrachloroethane)	*Nocardia rhodochrous*	0.63 - 0.99	[25]
androstanediol oxidation (in ethyl/butyl acetate)	co-immobilised dehydrogenases	0.36 - 0.8	[26]
progesterone hydroxylation (in oleic acid)	*Aspergillus ochraceous*	0.3 - 0.7	[27]
1- octene epoxidation (in n-hexadecane)	*Nocardia corrallina*	0.27 - 0.69	[28]
triglyceride hydrolysis (in isooctane)	lipase	0.1 - 0.8	[29]

TABLE 2

Reported values of phase ratio φ of some industrial applications

Reaction	Biocatalyst	Solvent	φ	Company	Ref
1-octene epoxidation	*Nocardia corallina*	n-hexadecane	0.5	Nippon-	[30]
styrene epoxidation		n-hexadecane	0.5	Mining	
1-tetradecane epoxidation		(reactant)	0.04		
phenol polymerisation	horseradish peroxidase	ethyl acetate	0.2	Mead	[31]
ester hydrolysis	subtilisin (immobilised)	dioxane	0.2	Bayer	[32]
		CHCl₃	0.2		
		1-butanol	0.2		
		MEK	0.5		
steroid dehydrogenation	*Arthrobacter simplex*	toluene	0.3	Upjohn	[33]
enantioselective ester hydrolysis	lipase	(reactant)	0.2-0.8*	Sumitomo	[34]

* φ was defined as % weight/volume

CONCLUSIONS

The selection of an appropriate phase ratio for operation of a two−liquid phase biocatalytic reactor depends on the characteristics and operational stability of the biocatalysts used. For instance, the phase ratio for optimal activity may not be optimal for catalyst stability. General guidelines are now available for the choice of phase ratio but much more needs to be done on engineering aspects of two−liquid phase systems before it will be possible to scale−up in a rational manner the many exciting biochemical conversions involving poorly water−soluble compounds now being reported.

REFERENCES

1. Smith, L.L., Steroids. In *Biotechnology*, vol. 6A, ed. K. Kieslich, Verlag Chemie, Basel, 1984, pp. 31-78.

2. Kloosterman, J. and Lilly, M.D., Effect of supersaturated aqueous hydrocortisone concentrations on the Δ¹-dehydrogenase activity of free and immobilized *Arthrobacter simplex*. *Enzyme Microb. Technol.*, 1984, 6, 113-6.

3. Dordick, J.S., Enzymatic catalysis in organic solvents. *Enzyme Microb. Technol.*, 1989, 11, 194-210.

4. Freeman, A. and Lilly, M.D., The effect of water-immiscible solvents on the Δ¹-dehydrogenase activity of free and PAAH-entrapped *Arthrobacter simplex*. *Appl. Microbiol. Biotechnol.*, 1987, 25, 495-501.

5. Lilly, M.D. and Woodley, J.M., Biocatalytic reactions involving water-insoluble organic compounds. In *Biocatalysts in Organic Syntheses*, eds J. Tramper, H.C. van der Plas and P. Linko, Elsevier, Amsterdam, 1985, pp. 179-92.

6. Lilly, M.D. and Dunnill, P., Use of immobilized biocatalysts for conversions of water-insoluble reactants: interesterification of fats. *Ann. N. Y. Acad. Sci.*, 1988, 501, 113-8.

7. Macrae, A.R., Interesterification of fats and oils. In *Biocatalysts in Organic Syntheses*, eds J. Tramper, H.C. van der Plas and P. Linko, Elsevier, Amsterdam, 1985, pp. 195-208.

8. Klibanov, A.M., Enzymes that work in organic solvent. *Chemtech*, 1986, 16, 354-9.

9. Woodley, J.M., Brazier, A.J. and Lilly, M.D., Lewis cell studies to determine reactor design data for two-liquid phase bacterial and enzymic reactions. *Biotechnol. Bioeng.*, Submitted for publication.

10. Lilly, M.D., Harbron, S. and Narendranathan, T.J., Two-liquid phase biocatalytic reactors. *Methods in Enzymology*, 1988, 136, 138-49.

11. Woodley, J.M. and Lilly, M.D., Extractive biocatalysis: the use of two-liquid phase biocatalytic reactors to assist product recovery. Submitted for presentation at ISCRE II.

12. Hanson, C., Mass transfer with simultaneous chemical reaction. In *Recent advances in liquid-liquid extraction*, ed. C. Hanson, Pergamon, Oxford, 1971, pp. 429-53.

13. Woodley, J.M., Cunnah, P.J. and Lilly, M.D., Stirred tank two-liquid phase biocatalytic reactor studies: kinetics, evaluation and modelling of substrate mass transfer. *Biocatalysis*, Submitted for publication.

14. Woodley, J.M., Stirred tank power input data for the scale-up of two-liquid phase biotransformations. Published in these Proceedings.

15. Laane, C., Boeren, S. and Vos, K., On optimizing organic solvents in multi-liquid-phase biocatalysis. *Trends Biotechnol.*, 1985, 3, 251-2.

16. Williams, A.C., Woodley, J.M., Ellis, P.A., Narendranathan, T.J. and Lilly, M.D., A comparison of pig liver esterase and *Bacillus subtilis* as catalysts for the hydrolysis of menthyl acetate in stirred two-liquid phase reactors. *Enzyme Microb. Technol.*, In press.

17. Williams, A.C., Woodley, J.M., Ellis, P.A. and Lilly, M.D., Denaturation and inhibition studies in a two-liquid phase biocatalytic reaction: the hydrolysis of menthyl acetate by pig liver esterase. In *Biocatalysis in Organic Media*, ed. C. Laane, J. Tramper and M.D. Lilly, Elsevier, Amsterdam, 1987, pp. 399-404.

18. Hocknull, M.D. and Lilly, M.D., Stability of the steroid Δ^1-dehydrogenation system of *Arthrobacter simplex* in organic solvent-water two-liquid phase environments. *Enzyme Microb. Technol.*, 1988, 10, 669-74.

19. Hocknull, M.D. and Lilly, M.D., The use of free and immobilised *Arthrobacter simplex* in organic solvent/aqueous two-liquid phase reactors. *Enzyme Microb. Technol.*, In press.

20. Lilly, M.D., Brazier, A.J., Hocknull, M.D., Williams, A.C. and Woodley, J.M., Biological conversions involving water-insoluble organic compounds. In *Biocatalysis in Organic Media*, ed. C. Laane, J. Tramper and M.D. Lilly, Elsevier, Amsterdam, 1987, pp. 3-17.

21. Jenkins, R.O., Stephens, G.M. and Dalton, H., Production of toluene *cis*-glycol by *Pseudomonas putida* in glucose fed-batch culture. *Biotechnol. Bioeng.*, 1987, 29, 873-83.

22. Brazier, A.J., Lilly, M.D. and Herbert, A.B., Toluene *cis*-glycol synthesis by *Pseudomonas putida*; kinetic data for reactor evaluation. *Enzyme Microb. Technol.*, In press.

23. Brookes, I.K., Lilly, M.D. and Drozd, J.W., Stereospecific hydrolysis of *d,l*-menthyl acetate by *Bacillus subtilis*: mass transfer-reaction interactions in a liquid-liquid system. *Enzyme Microb. Technol.*, 1986, 8, 53-7.

24. Harbron, S., Smith, B.W. and Lilly, M.D., Two-liquid phase biocatalysis: epoxidation of 1,7-octadiene by *Pseudomonas putida*. *Enzyme Microb. Technol.*, 1986, 8, 85-8.

25. Duarte, J.M.C., The enzymic oxidation of cholesterol in the presence of water-immiscible solvents. Ph.D. thesis, University of London, 1982.

26. Carrea, G., Riva, S., Bovara, R. and Pasta, P., Enzymatic oxido-reduction of steroids in two-phase systems: effects of organic solvents on enzyme kinetics and evaluation of the performance of different reactors. *Enzyme Microb. Technol.*, 1988, 10, 333-40.

27. Ceen, E.G., Herrmann, J.P.R. and Dunnill, P., Two-liquid phase reactor studies of 11α-hydroxylation of progesterone by *Aspergillus ochraceus*. *Biotechnol. Bioeng.*, 1988, 31, 743-6.

28. Kawakami, K., Characterizaton of production of 1,2-epoxyoctane from 1-octene by non-growing cells of *Nocardia corallina* B-276 in aqueous-organic media. Presented at Enzyme Engineering X Conference, Kashikojima, Japan, September 1989.

29. Mukataka, S., Kobayashi, T. and Takahashi, J., Kinetics of enzymatic hydrolysis of lipids in biphasic organic-aqueous systems. *J. Ferment. Technol.*, 1985, 63, 461-6.

30. Furuhashi, K., Shintani, M. and Takagi, M., Effects of solvents on the production of epoxides by *Nocardia corallina* B-276. *Appl. Microbiol. Biotechnol.*, 1986, 23, 218-23.

31. Pokora, A.R. and Cyrus, W.L., Phenolic developer resins. U.S. patent No. 4,647,952, 1987.

32. Schutt, H., Schmidt-Kastner, G., Arens, A. and Preiss, M., Preparation of optically acitve D-arylglycines for use as side chains for semi-synthetic penicillins and cephalosporins using immobilized subtilisins in two-phase systems. *Biotechnol. Bioeng.*, 1985, 27, 420-33.

33. Evans, T.W., Kominek, L.A., Wolf, H.J. and Henderson, S.L., Steroid dehydrogenation. Eur. patent appl. No. 0 127 294 A1.

34. Umemura, T. and Hirohara, H., Preparation of optically active pyrethroids via enzymatic resolution. In *Biocatalysis in Agricultural Biotechnology*, eds J.R. Whitaker and P.E. Sonnet, Amer. Chem. Soc., Washington, 1989, pp.371-84.

APPLICATION OF LIPASES FOR SYNTHESIS
OF NEW CHEMICALS

by

Sven Erik Godtfredsen
Novo Nordisk
DK-2880 Bagsvaerd
Denmark

ABSTRACT

So far, the use of biocatalysts in organic chemical processing has acquired limited importance only. It has proven exceedingly difficult to introduce biocatalysts into well optimized chemical processes in an economical fashion. In regard to new classes of chemicals which are of no importance today due to difficulties in their manufacturing by classical methods the biocatalysts may, however, gain importance. An example of such future chemicals may be sugar esters now amenable for preparation in an economical fashion with the aid of biocatalysts.

During the recent years the interest for application of enzymes in organic chemical processing has been increasing steadily. The possibility of replacing polluting chemical processes with environmentally more acceptable enzymatic procedures is an important reason behind this interest in enzymatic synthesis. Also important for the increasing interest is the possibility of exploiting the unique properties of the enzymes as e.g. their regio- and stereoselectivity for synthesis of organic chemicals in an economical competitive fashion. Finally, the increasing availability of new industrial enzymes is an important factor behind the interest of the organic chemists in the use of biocatalysts.
Even so, the use of enzymes in organic chemical processing is, so far, of very limited importance when viewed in the perspective of the entire chemical industry. Very few enzyme based processes only are currently being applied industrially and in spite of the fact that new synthetic applications of enzymes are published daily very few new processes are actually being exploited in practise.

An important reason behind this apparent lack of success of enzymatic synthesis has to do with the close connection between the scope of classical organic chemical methods and the nature of chemicals successfully developed. Those chemicals which, today, enjoy a big market are thus those which are indeed amenable for production by classical means and which are, accordingly, produced in established plants based on highly optimized classical methods. It will in most instances be very difficult, if not impossible, to develop enzymatic processes which can compete with these highly competitive processes, even more so because the compounds thus produced are not, in most instances, amenable to enzymatic synthesis.

In our strategy for development of enzymes for organic chemical processing we have, as a consequence of these considerations, chosen to concentrated on development of enzymatic syntheses of chemicals which have the potential of being developed into commodity chemicals and which are particularly amenable for enzymatic synthesis while not readily available by conventional chemical methods.

A specific example of a class of compounds we have considered in this connection comprise surface active materials composed of carbohydrates and fatty acids. Such compounds are green chemicals in the sense that they can be synthesized from readily available agrochemical products, that they are degraded readily in the environment to natural sugars and fats[1], and that they might be prepared with the aid of enzymes in "green" processes. Provided such compounds could be made available at a low cost they might for all these reasons gain industrial importance.

Before us many workers had attempted to couple carbohydrates and fatty acids enzymatically[2], however without success: yields reported are low, consumption of enzymes unrealistically high, and the process conditions outside the industrially feasible range[3, 4].

We have overcome the difficulties encountered by other workers by employing glycosides instead of free sugars in lipase catalysed synthesis of surfactant esters. In this fashion we have developed a highly efficient lipase catalyzed process for regiospecific esterification of the primary hydroxyl group in simple alkyl glycosides, (Scheme). This procedure allows us e.g. to esterify ethyl (D)-glucopyranoside with C_8-C_{18} fatty acids in yields of 85-95 % of the 6-O-monoesters using an immobilized lipase from a species of _Candida antarctica_[5] as catalyst (Table 1). The reaction was performed very conveniently simply by mixing the two reactants at 70 °C under reduced pressure in the presence of the thermostable immobilised lipase. The purity of the glucoside esters thus obtained is excellent. The surfactant properties have been found to be similar to those of common nonionic surfactants (Table 1).

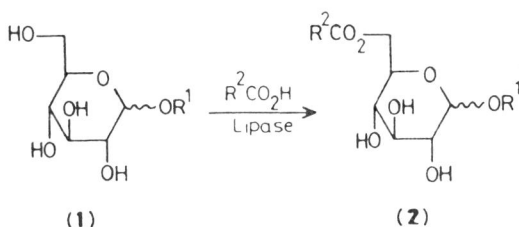

R^1 = H, Me, Et, Pr^n, Pr^i, or Bu^n. R^2 = C_7H_{15}, C_9H_{19}, $C_{11}H_{23}$, $C_{13}H_{27}$, $C_{15}H_{31}$, or $C_{17}H_{35}$.

Scheme 1

Table 1. Yield, critical micelle concentration (CMC) and surface tension of fatty acid esters of ethyl (D)-glucopyranoside.[a])

Fatty acid	Yield (%)	CMC (mol/l)	Min/dyn·cm^{-1}
Octanoic acid	86.9	$2.0 \ 10^{-3}$	31
Decanoic acid	88.4	$9.6 \ 10^{-4}$	31
Dodecanoic acid	85.8	$5.1 \ 10^{-5}$	31
Tetradecanoic acid	89.1	$5.3 \ 10^{-5}$	33
Hexadecanoic acid	93.1	$1.8 \ 10^{-4}$	39
Octadecanoic acid	95.5	$8.3 \ 10^{-6}$	44
Octadecenoic acid	91.8	$3.4 \ 10^{-4}$	35

[a]) The standard reaction was performed by mixing 50 g (0.24 mol) of ethyl (D)-glucopyranoside at 70 °C and 0.01 bar, with 1.35 equivalent of fatty acid and 2.5 % (w/w) of immobilized Candida antarctica lipase of an activity of 40 BIU/g[#]. The yields indicated refer to yields of product obtained after purification by chromatography on silica gel using a gradient of petroleum ether, ethyl acetate and methanol as eluent. The CMC and surface tension were determined using a Krüss tensiometer type K 10. MS and NMR (^{1}H- and ^{13}C-NMR) data corresponding to the expected reaction products were obtained for all compounds synthesized.

Previous attempts by other groups to enzymatically esterify α-(D)-glucose and methyl α-(D)-glucopyranosides have not been successful.[4] As indicated, however, we found a dramatic change in reactivity of ethyl (D)-glucopyranose as compared to either glucose or methyl α-(D)-glucopyranoside (Table 2). Presumably, the higher reactivity of the glucosides are due to higher solubilities of the reactants in one another and to the substrate selectivity of the enzyme used. Our best yields were obtained using C_8-C_{18} fatty acids and ethyl or isopropyl glucoside as substrates and, as catalyst, a heat stable lipase derived from a strain of Candida antarctica (Table 1, 2 and 3). This particular enzyme is non-specific in regard to triglyceride hydrolysis reactions but exhibits a very high selectivity towards the primary hydroxy group in the synthesis reactions described.

Table 2. Rate of reaction of selected carbohydrates with dodecanoic acid.[b])

Carbohydrate	T 1/2[c])	Conversion 6h	24h	%Di-ester
α-(D)-glucose	> 1 week	–	< 5%	–
Methyl α-(D)-glucopyranoside	22 h	20.0	53.3	3.5
Ethyl (D)-glucopyranoside	2.5 h	74.0	92.5	4.9
Isopropyl (D)-glucopyranoside	2.1 h	70.0	93.2	4.2
n-Propyl (D)-glucopyranoside	1.4 h	79.2	95.6	17.3
n-Butyl (D)-glucopyranoside	1.0 h	78.5	94.4	21.8

[b]) Conversion of different carbohydrates was performed at 70 $^{\circ}$C and 0.01 bar, using 1.5 equivalent of dodecanoic acid and 6 % (w/w) of immobilized Candida antarctica lipase of 40 BIU/g[#]. The progress of the reactions were monitored by HPLC.

[c]) Time for conversion of 50% of the starting carbohydrate derivatives.

The esterification reaction was found to be catalyzed by a variety of enzymes, however with quite different conversion rates and selectivities (Table 3). Lipozyme[TM], an immobilized Mucor miehei lipase is a 1,3-specific lipase with respect to hydrolysis of triglycerides and as such preferentially catalyses reactions with primary hydroxy groups, e.g. in acidolysis reactions explored commercially using this enzyme. Even so, the activity and selectivity of this enzyme was lower than that of the unspecific Candida lipase when expressed in synthesis of glucoside esters. To achieve the best enzyme activity the immobilized enzyme preparations were moisturized to a water content of about 10 %.[6]

The yields obtained using short chain fatty acids, C_8-C_{10}, were generally lower than those realized using their higher homologues (Table 1). This may be due to an increased solubility in water of the short chain fatty acids which cause a pH change in the water bound to the enzyme and which may also dissolve some water from the enzyme surface and thereby cause a loss of activity.[7] The enzymatic synthesis of glycoside esters has been tested at pilot plant level and proved applicable for synthesis of kg-quantities of the esters. We expect that the method can also be adapted to production of actual industrial quantities of the glycolipids which may, therefore, become commodity chemicals in the future and contribute to the success of enzymatic synthesis.

Table 3. Lipase activity, rate of conversion and diester content in the synthesis using different immobilized lipases.[d]

Lipase from	After 24 h			After 48 h	
	Activity BIU/g[#]	Conv. (%)	Diester (%)	Conv. (%)	Diester (%)
Candida antarctica	40	96.1	3.1	96.3	5.2
Mucor miehei	25	97.9	19.3	97.9	28.5
Humicola sp.	69	99.0	45.8	99.7	64.3
Candida cylindracea	29	21.2	1.2	39.6	6.1
Pseudomonas sp.	181	1.5	–	14.3	21.6

[d] The conversion of ethyl (D)-glucopyranoside was performed at 70 °C and 0.01 bar, using 2 equivalents of dodecanoic acid and 6 % (w/w) of different immobilized lipases. The conversions were measured by HPLC.

One Batch Interesterification Unit (BIU) corresponds to 1 μmol of hexadecanoic acid incorporated (initial activity) into trioctadecenoyl glycerol per minute.

& All glucosides except methyl α-(D)-glucopyranoside were mixtures of anomers obtained by glucosidation of the desired alcohol with α-(D)-glucose.[8]

REFERENCES

1. H. Baumann, M. Bühler, H. Fochem, F. Hirsinger, H. Zoebelein, and J. Falbe, Angew. Chem. Int. Ed. Engl., 1988, 27, 41.

2. E. Reinefeld, and H-F. Korn, Die Stärke, 1968, 20, 181.
K. Yoshimoto, K. Tahara, S. Suzuki, K. Sasaki, Y. Nishikawa, and Y. Tsuda, Chem. Pharm. Bull., 1979, 27, 2661.
E. Albano-Garcia, E.G. Lorica, M. Pama, and L. de Leon, Philipp. J. Coconut Stud., 1980, 5, 51.
D. Plusquellec, and K. Baczko, Tetrahedron Lett., 1987, 28, 41.
D.V. Myhre US pat. 3,597,417, 1971.
G.N. Bollenback, F.W. Parrish, Carbohydr. Res., 1971, 17, 431.

3. J.M. Sugihara, Adv. Carbohydr. Res., 1953, 8, 1.

4. M. Therisod, and A.M. Klibanov, J. Amer. Chem. Soc., 1986, 108, 5638.
H.M. Sweers, and C-H. Wong, J. Amer. Chem. Soc., 1986, 108, 6421.
J. Chopineau, F.D. McCafferty, M. Therisod, and A.M. Klibanov, Biotechnol. Bioeng., 1988, 31, 208.
S. Riva, J. Chopineau, A.P.G. Kieboom, and A.M. Klibanov, J. Amer. Chem. Soc., 1988, 110, 584.
J-F. Shaw, and A.M. Klibanov, Biotechnol. Bioeng., 1987, 29, 648.
M. Therisod, and A.M. Klibanov, J. Amer. Chem. Soc., 1987, 109, 3977.

M. Kloosterman, E.W.J. Mosmuller, H.E. Schoemaker, and E.M. Meijer, <u>Tetrahedron Lett.</u>, 1987, <u>28</u>, 2989.

5. H.P. Heldt-Hansen, I. Michiyo, S.A. Patkar, T.T. Hansen, P. Eigtved, 1988, Proc. Symposium on Biocatalysis and Biomimetics: Aspects of Enzyme Chemistry for Agriculture-Enzyme Production of Speciality Products, in press.

6. A. Zaks, and A.M. Klibanov <u>J. Biol. Chem.</u>, 1988, <u>263</u>, 8017.

7. C. Laane, S. Boeren, K. Vos, and C. Veeger, <u>Biotechnol. Bioeng.</u>, 1987, <u>30</u>, 81.
M. Norin, J. Boutelje, E. Holmberg, and K. Hult, <u>Appl. Microbiol. Biotechnol.</u>, 1988, <u>28</u>, 527.

8. E. Fischer, and L. Bensch, <u>Ber.</u>, 1894, <u>27</u>, 2478.
J.E. Cadotte, F. Smith, and D. Spriesterbach, <u>J. Amer. Chem. Soc.</u>, 1952, <u>74</u>, 1501.

STEROID TRANSFORMATION BY RECOMBINANT YEAST CELLS EXPRESSING P450 MONOOXYGENASES

HIDEO OHKAWA, YOSHIYASU YABUSAKI, TOSHIYUKI SAKAKI,
HIROKO MURAKAMI AND MEGUMI SHIBATA
Biotechnology Laboratory, Takarazuka Research Center,
Sumitomo Chemical Co., Ltd.
Takarazuka, Hyogo 665, Japan

ABSTRACT

Microsomal P450 monooxygenase enzymes, P45017α and P450C21, which are involved in the steroidogenesis in bovine adrenal cortex, were expressed in the yeast Saccharomyces cerevisiae. The recombinant yeast strains specifically converted substrates into the corresponding 17α-hydroxylation (P45017α) and C21-hydroxylation (P450C21) products. The strains expressing P45017α/yeast reductase and P450C21/yeast reductase fused enzymes showed highest hydroxylase activities. These recombinant strains seem to have a potential for bioconversion of steroid compounds through biosynthesis pathways.

INTRODUCTION

Microsomal P450 monooxygenases, which consist of a specific cytochrome P450 (P450) and a generic NADPH-cytochrome P450 reductase (reductase), metabolize a wide variety of lipophilic compounds including endogenous substrates such as steroid hormones and fatty acids as well as foreign chemicals such as drugs and environmental chemicals. The metabolic versatility of P450 monooxygenases can be explained in part by the multiplicity of P450 species. For example, rat liver microsomes contain over ten P450 species, each of which recieves electrons from NADPH by the catalysis of the same reductase, and introduces an activated oxygen into a substrate, yielding an hydroxylation product. Although hydroxylation reactions are important for the production of various industrial chemicals, P450 monooxygenases were so far thought not to be useful for industrial purposes, because of complexity of the enzyme system, difficult handling of the membrane-bound enzymes, and low turnover rates. So, we attempted to simplify the enzyme system, and to increase the reactivity and stability of the enzymes by genetic engineering technology.

YEAST EXPRESSION SYSTEM

We established a yeast expression system for cDNAs encoding microsomal P450 monooxygenase enzymes, as shown in Fig. 1 [1]. An expression plasmid was constructed by inserting a P450 cDNA between the yeast alcohol dehydrogenase I (ADH) promoter and terminator of the expression vector pAAH5. The yeast Saccharomyces cerevisiae AH22 cells were transformed with the expression plasmid by LiCl method. The transformed AH22 cells produced the active enzyme, inserted it into microsomal membranes and exhibited P450-dependent monooxygenase activity by coupling with endogenous yeast reductase. No particular genetic information was needed for the incorporation of heme into the enzyme. Thus, the monooxygenase activity of the transformed yeast cells was easily assayed by measuring an amount of the hydroxylation product in the culture supernatant using TLC or HPLC after incubation with the substrate, because the cells took up the substrate, hydroxylated it and released the product into the medium. Also, the cellular P450 content was measured by the reduced CO-difference spectrum of the whole yeast cells.

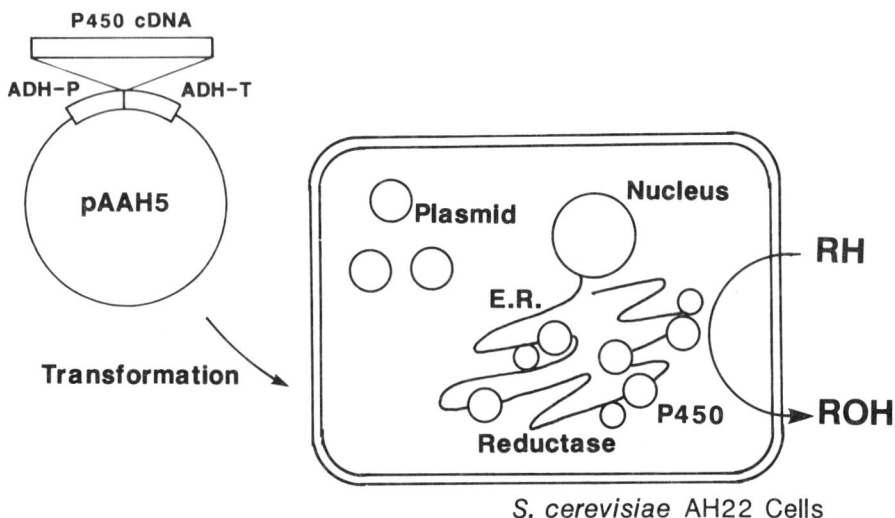

FIGURE 1. A yeast expression system for microsomal P450 cDNAs.

With the yeast expression system, we succeeded in the functional expression of a number of P450 monooxygenase enzymes; rat P450c [2], chimeric P450s consisting of rat P450c and P450d [3], rat reductase [4], and N-terminal-truncated enzymes of P450c and reductase [5]. Thus, the yeast expression system was found to be suitable for the functional expression of microsomal P450 monooxygenase enzymes. In addition, we constructed rat P450c/reductase fused enzymes by the expression in the yeast of the fused genes of P450c cDNA with the DNA fragment for the N-terminal-truncated reductase [6]. The fused enzymes showed the enhanced hydroxylase activity, probably due to efficient electron transfer through intramolecular interaction between the P450 and reductase moieties (Fig. 2). The presence of the N-terminal anchor region of the reductase adversely affected both expression level and enzyme activity of the fused enzymes [5].

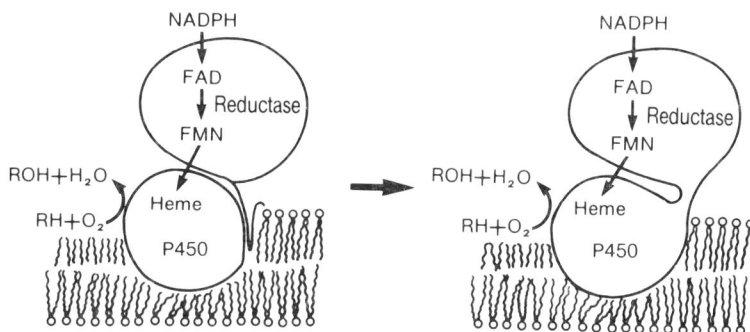

FIGURE 2. Construction of a P450/reductase fused enzyme.

STEROIDOGENIC P450

Steroidogenesis from cholesterol to cortisol in adrenal cortex involves hydroxylation reactions catalyzed by four distinct P450 species; P450SCC, P45017α, P450C21 and P45011β. Of these, P45017α and P450C21 are microsomal enzymes, whereas P450SCC and P45011β occur in mitochondria. P45017α catalyzes 17α-hydroxylation toward pregnenolone and progesterone to yield the corresponding 17α-hydroxylation products. P450C21 mediates C21-hydroxylation toward progesterone and 17α-hydroxyprogesterone to give the corresponding C21-hydroxylation products. These products are important intermediates for glucocorticoid production. So, we attempted to express bovine P45017α and P450C21 cDNAs in the yeast to create recombinant yeast strains, which are useful for bioconversion of steroid compounds. Both P45017α and P450C21 cDNAs were isolated from a bovine adrenal cDNA library in λgt11 by screening with synthetic DNA probes. Also, yeast reductase cDNA and genomic DNA were cloned from S. cerevisiae [7].

P45017α [8,9]

    The expression plasmids pAα2, pARα1 and pαLY3 were constructed for P45017α, both P45017α and yeast reductase, and a P45017α/yeast reductase fused enzyme, respectively (Fig. 3). The P45017α cDNA was under the control of ADH promoter and terminator. The yeast reductase gene in pARα1 had its own promoter and terminator. The hybrid cDNA encoding the fused enzyme was constructed by connecting the P45017α cDNA with the DNA fragment for yeast reductase lacking the N-terminal hydrophobic 41 amino acid residues, which corresponded to the membrane anchor.

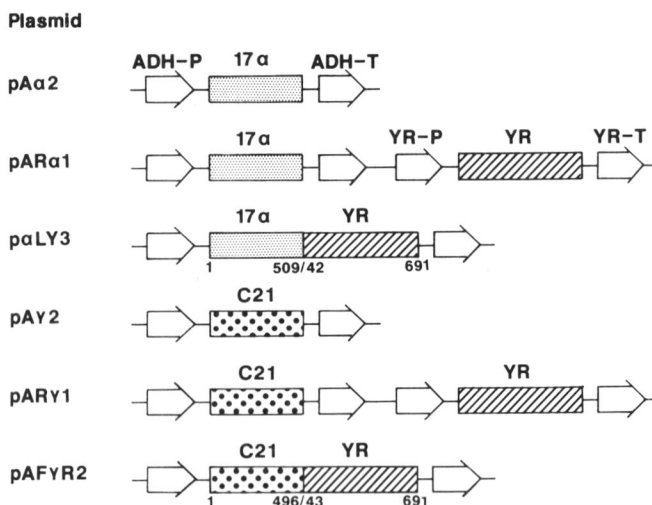

FIGURE 3. Structures of expression plasmids.
         17α; P45017α, C21; P450C21, YR; yeast reductase,
         ADH-P; ADH promoter, ADH-T; ADH terminator,
         YR-P; yeast reductase promoter, YR-T; yeast
         reductase terminator.

    On incubation with ^{3}H-labeled progesterone, the culture of the transformed AH22/pAα2 cells expressing P45017α specifically converted the substrate into 17α-hydroxyprogesterone, as confirmed by TLC analysis. The transformed AH22/pARα1 cells expressing simultaneously both P45017α and yeast reductase hydroxylated the substrate to yield the 17α-hydroxylation product at a faster rate than the AH22/pAα2 cells. Both strains contained nearly the same amounts of P45017α as determined by the reduced CO-difference spectra, while the AH22/pARα1 strain contained a much higher amount of the reductase than the AH22/pAα2 strain. The specific 17α-hydroxylase activity was five times higher in the AH22/pARα1 strain than that in the AH22/pAα2 strain (Table 1). Thus, it is likely that bovine P45017α interacts with endogenous yeast reductase on the microsomal membrane to exhibit the 17α-hydroxylase activity, and the overproduction of yeast reductase enhanced the 17α-hydroxylase activity in the yeast cells.

The transformed AH22/pαLY3 cells were confirmed to produce the P45017α/yeast reductase fused enzyme by Western blot using anti-P45017α and anti-yeast reductase IgGs. On incubation with ^{3}H-labeled progesterone, the culture of the AH22/pαLY3 cells specifically converted the substrate into 17α-hydroxyprogesterone. The reaction was as fast as that of the AH22/pARα1 strain. Although the hemoprotein content in the AH22/pαLY3 cells was slightly lower than the P45017α level in the AH22/pAα2 cells, the specific 17α-hydroxylase activity of the AH22/pαLY3 strain was seven times higher than that of the AH22/pAα2 strain. These results indicated that the P45017α/yeast reductase fused enzyme shows the enhanced 17α-hydroxylase activity.

TABLE 1

Hemoprotein contents and hydroxylase activities in the recombinant yeast strains

Strain	Enzyme	Hemoprotein ($\times 10^4$ molec./cell)	Hydroxylase (mol/min/ mol P450)	Relative activity (/ml culture)
AH22/pAα2	17α	13	9[a]	12
AH22/pARα1	17α, YR	11	43[a]	49
AH22/pαLY3	17α/YR	8	61[a]	50
AH22/pAγ2	C21	0.3	33[b]	1
AH22/pARγ1	C21, YR	0.2	150[b]	3
AH22/pAFγR2	C21/YR	1	200[b]	20

[a]17α-hydroxylase activity toward progesterone.
[b]C21-hydroxylase activity toward 17α-hydroxyprogesterone.

P450C21 [10]

The expression plasmids pAγ2, pARγ1 and pAFγR2 were also constructed for P450C21, both P450C21 and yeast reductase, and a P450C21/yeast reductase fused enzyme, respectively (Fig. 3). The fused enzyme encoded by pAFγR2 also lacked the N-terminal 42 amino acid residues of yeast reductase. On incubation with

17α-hydroxyprogesterone, the transformed AH22/pAγ2 cells expressing P450C21 converted the substrate into 11-deoxycortisol. The transformed AH22/pARγ1 cells expressing simultaneously both P450C21 and yeast reductase showed a three times higher C21-hydroxylase activity than the AH22/pAγ2 cells, although the P450C21 content was nearly the same in both strains. The specific C21-hydroxylase activity was five times higher in the AH22/pARγ1 strain than that in the AH22/pAγ2 strain (Table 1). These results indicated that the overproduction of yeast reductase increased the C21-hydroxylase activity in the yeast cells.

The transformed AH22/pAFγR2 cells expressing the P450C21/yeast reductase fused enzyme, which was confirmed by Western blot using anti-P450C21 and anti-yeast reductase IgGs, converted 17α-hydroxyprogesterone into 11-deoxycortisol at a faster rate than did the AH22/pAγ2 cells. The fused enzyme level in the AH22/pAFγR2 cells was three times higher than the P450C21 level in the AH22/pAγ2 cells. Also, the specific C21-hydroxylase activity was six times higher in the AH22/pAFγR2 strain than that in the AH22/pAγ2 strain (Table 1). So, it was found that the P450C21/yeast reductase fused enzyme was increased in both expression level and C21-hydroxylase activity in the yeast.

Both P45017α and P450C21

The plasmid pARαγ was constructed for simultaneous expression of P45017α, P450C21 and yeast reductase in the yeast (Fig. 4). Both P45017α and P450C21 cDNAs were each under the control of ADH promoter and terminator, and the reductase gene had its own promoter and terminator. The transformed AH22/pARαγ cells were assayed for both 17α-hydroxylase and C21-hydroxylase activities. After addition of ^{3}H-labeled progesterone, the recombinant cells rapidly converted the substrate into 17α-hydroxyprogesterone, which was then converted into 11-deoxycortisol, although other metabolites were also detected by TLC (Fig. 5). The conversion rate of progesterone was about 90% and that of 17α-hydroxyprogesterone was 60%, because the P450C21 level was fairly lower than the P45017α level in the transformed yeast. Thus, it was found that the recombinant yeast cells simultaneously expressing P45017α, P450C21 and yeast reductase converted progesterone to 11-deoxycortisol through 17α-hydroxylation and C21-hydroxylation.

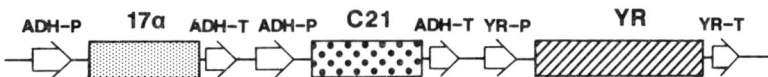

FIGURE 4. The plasmid pARαγ for simultaneous expression of P45017α, P450C21 and yeast reductase.

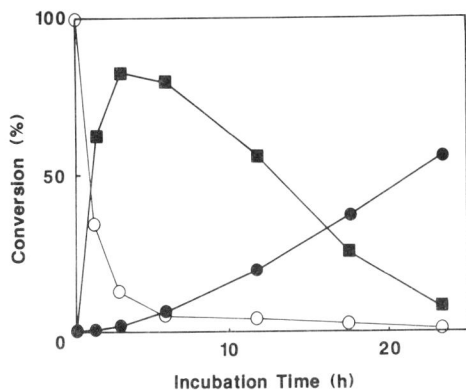

FIGURE 5.
Conversion of progesterone
into 11-deoxycortisol by
the AH22/pARαγ strain.
O : progesterone
■ : 17α-hydroxyprogesterone
● : 11-deoxycortisol

DISCUSSION

Among three strains expressing P45017α and its fused enzyme,
the expression level of hemoproteins was nearly the same.
However, the specific 17α-hydroxylase activity was the highest
in the strain expressing the fused enzyme. So, this strain
seems to have a potential for bioconversion of progesterone
into 17α-hydroxyprogesterone. Also, with three strains
expressing P450C21 and its fused enzyme, both expression level
and specific C21-hydroxylase activity were the highest in the
strain expressing the fused enzyme, which is a promising
strain for bioconversion of 17α-hydroxyprogesterone into 11-
deoxycortisol. The combined use of both recombinant strains
possibly converts progesterone into 11-deoxycortisol through
17α-hydroxylation and C21-hydroxylation (Fig. 6). 11-
Deoxycortisol can be converted into cortisol by 11β-
hydroxylation. These are important intermediates for
glucocorticoid production.

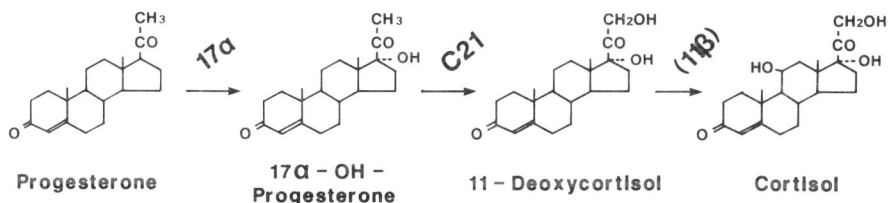

FIGURE 6. Possible bioconversion processes for steroid
compounds.

The bioconversion processes using the recombinant yeast strains make it possible to shorten ordinary processes for steroid compound production, although further studies are needed for practical use.

ACKNOWLEDGEMENT

This work was performed as a part of the Research and Development Project of Basic Technologies for Future Industries supported by NEDO (New Energy and Industrial Technology Development Organization).

REFFERENCES

1. Yabusaki, Y. and Ohkawa, H., Genetic engineering on P-450 monooxygenases. In Frontiers in Biotransformation Vol. 4, ed. K. Ruckpaul, Akademie-Verlag, Berlin and Taylor & Francis, London, 1990, in press.

2. Oeda, K., Sakaki, T. and Ohkawa, H., Expression of rat liver cytochrome P450MC cDNA in Saccharomyces cerevisiae. DNA, 1985, 4, 203-210.

3. Sakaki, T., Shibata, M., Yabusaki, Y. and Ohkawa, H., Expression in Saccharomyces cerevisiae of chimeric cytochrome P450 cDNAs constructed from cDNAs for rat cytochrome P450c and P450d. DNA, 1987, 6, 31-39.

4. Murakami, H., Yabusaki, Y. and Ohkawa, H., Expression of rat NADPH-cytochrome P450 reductase cDNA in Saccharomyces cerevisiae. DNA, 1986, 5, 1-10.

5. Yabusaki, Y., Murakami, H., Sakaki, T., Shibata, M and Ohkawa, H., Genetically engineered modification of P450 monooxygenases: functional analysis of the amino-terminal hydrophobic region and hinge region of the P450/reductase fused enzyme. DNA, 1988, 7, 701-711.

6. Murakami, H., Yabusaki, Y., Sakaki, T., Shibata, M and Ohkawa, H., A genetically engineered P450 monooxygenase: construction of the functional fused enzyme between rat cytochrome P450c and NADPH-cytochrome P450 reductase. DNA, 1987, 6, 189-197.

7. Yabusaki, Y., Murakami, H. and Ohkawa, H., Primary structure of Saccharomyces cerevisiae NADPH-cytochrome P450 reductase deduced from nucleotide sequence of its cloned gene. J. Biochem., 1988, 103, 1004-1010.

8. Sakaki, T., Shibata, M., Yabusaki, Y., Murakami, H. and Ohkawa, H., Expression of bovine cytochrome P450c17 cDNA in Saccharomyces cerevisiae. DNA, 1989, 8, 409-418.

9. Shibata, M., Sakaki, T., Yabusaki, Y., Murakami, H. and
   Ohkawa, H., Genetically engineered P450 monooxygenases;
   construction of bovine P450c17/yeast reductase fused
   enzymes. DNA, 1990, in press.

10. Sakaki, T., Shibata, M., Yabusaki, Y., Murakami, H. and
    Ohkawa, H., Genetically engineered construction of bovine
    cytochrome P450c21/yeast reductase fused enzymes in
    Saccharomyces cerevisiae., in preparation.

IDENTIFICATION AND DESIGN OF BINDING DETERMINANTS IN PROTEINS

James A. Wells and Brian C. Cunningham
Department of Biomolecular Chemistry
Genentech, Inc., 460 Pt. San Bruno Blvd.
South San Francisco, CA 94080 U.S.A.

ABSTRACT

Mutagenic analyses are described for probing and designing binding determinants of two different proteins. The first protein is an enzyme, subtilisin, for which structural models were available for the enzyme substrate complex. The second protein is human growth hormone for which a structural model of its complex with receptor was **not** available. In both cases, it was possible to identify and engineer binding determinants that lead to large changes in affinity.

INTRODUCTION

Homologous proteins can vary widely in functional properties, especially binding specificity. This is perhaps most evident in the family of mammalian serine proteases where homologous enzymes such as trypsin, elastase and chymotrypsin exhibit dramatically different substrate specificities (for review see Ref. 1). Comparisons of their X-ray elucidated structures revealed important structural differences between these enzymes in the residues that contact the substrate.

RESULTS AND DISCUSSION

To test the importance of residues making direct contact with the substrate we have engineered the substrate binding site of a serine protease, subtilisin, that is convergently related to the mammalian family of enzymes (for review see Ref. 2). Mutations within the substrate binding site that create changes in electrostatic (3), steric and hydrophobic interactions (4), cause large (and somewhat predictable) changes in substrate specificity.

It is also possible to recruit the substrate specificity profile from one natural variant subtilisin (from *Bacillus amyloliquefaciens*) into another (*B. licheniformis*) by changing only residues that can make direct contact with the substrate (5). These studies show that engineering contact residues identified by high resolution structural analysis can have huge effects on substrate specificity.

It is often, in fact usually the case, that for a protein of interest one does not have a high resolution structure of the protein-ligand complex. This was precisely our situation when faced with the problem of locating the receptor binding determinants in human growth hormone (hGH). This was important in order to determine the feasibility of designing a small molecule analog of this hormone.

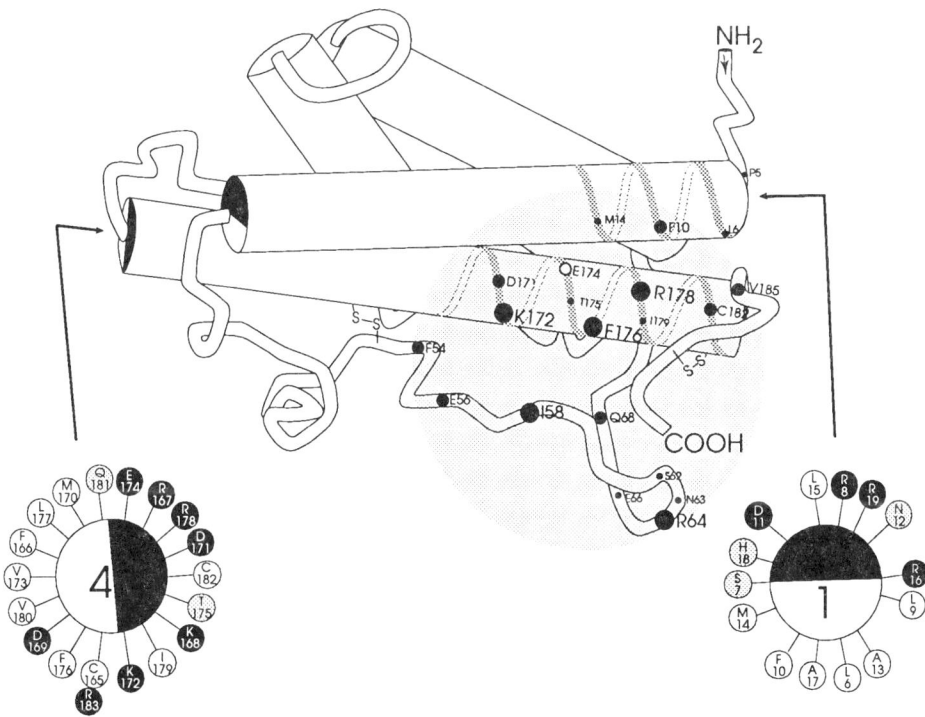

Figure 1. Location of residues in hGH that strongly modulate its binding to the hGH binding protein. Alanine substitutions (serine or asparagine in the case of T175 or R178, respectively) are indicated that cause >10-fold reduction (●), a 4- to 10-fold reduction (◉) or increase (○), or a 2- to 4-fold reduction (•), in binding affinity (7). Helical wheel projections in regions of α-helix reveal their amphipathic quality. Note that in helix 4 the most important determinants are on its hydrophilic face (shaded). The structural model (6) was derived from a folding diagram of porcine GH determined crystallographically (9). Figure is taken from reference 8.

Two complementary mutagenic strategies have been employed to define residues in hGH that are most important for binding to the cloned hGH receptor from liver. In the first approach termed homolog-scanning mutagenesis (6), segments (7 to 30 residues long) of non-binding homologs (human prolactin or placental lactogen) were substituted throughout the hGH molecules (191 residues in length). Receptor binding was dramatically disrupted by a small set of these segment substituted molecules that mapped primarily to two separate regions of the hGH sequence; these regions formed a patch when mapped on to a low resolution model derived from porcine growth hormone.

A higher-resolution strategy termed alanine-scanning mutagenesis was then used to identify specific side-chains within the binding patch that were most important in binding. From 62 single alanine mutants, twelve side-chains were identified that caused large reductions in binding affinity (4- to 20-fold); one mutant E174A was found that caused more than a 4-fold increase in affinity. These two systematic approaches produced a high-resolution functional map of residues important for receptor (Fig. 1) or antibody interactions (6,7). Recently, this information has been used to incorporate the receptor binding determinants into human prolactin to recruit it to bind to the hGH receptor (8).

In summary, protein residues in direct contact with a ligand are extremely important in conferring binding affinity. These may be localized by direct structural analysis or inferred from mutagenic analysis. By either method of identification, when such binding determinants from homolog X are introduced into another (homolog Y) which may have a dramatically different ligand specificity, it is possible to recruit the specificity property of homolog X into Y. This is a useful protein design strategy because it allows one to combine the most useful functional properties (such as specificity, catalysis, stability, etc.) contained within the family into one homolog.

REFERENCES

1. Kraut, J., Serine proteases: structure and mechanism of catalysis. *Annu. Rev. Biochem.*, 1977, **46**, 331-358.

2. Wells, J.A. and Estell, D.A., Subtilisin - an enzyme designed to be engineered, *Trends in Biochem. Sci.*, 1988, **13**, 291-297.

3. Wells, J.A., Powers, D.B., Bott, R.R., Graycar, T.P., and Estell, D.A. Designing substrate specificity by protein engineering of electrostatic interactions. *Proc. Natl. Acad. Sci. USA*, 1987, **84**, 1219-1223.

4. Estell, D.A., Graycar, T.P., Miller, J.V., Powers, D.B., Burnier, J.P., Ng, P.G., and Wells, J.A., Probing steric and hydrophobic effects on enzyme substrate interactions by protein engineering. *Science*, 1986, **233**, 659-663.

5. Wells, J.A., Cunningham, B.C., Graycar, T.P., and Estell, D.A. Recruitment of substrate-specificity properties from one enzyme into a related one by protein engineering, *Proc. Natl. Acad. Sci. USA*, 1987, **84**, 5167-5171.

6. Cunningham, B.C., Jhurani, P., Ng, P., and Wells, J.A. Receptor and antibody epitopes in human growth hormone identified by homolog-scanning mutagenesis. *Science*, 1989, **243**, 1330-1336.

7. Cunningham, B.C. and Wells, J.A. High-resolution epitope mapping of hGH-receptor interactions by alanine-scanning mutagenesis. *Science*, 1989, **244**, 1081-1085.

8. Cunningham, B.C., Henner, D.J., and Wells, J.A. Engineering human prolactin to bind to the human growth hormone receptor. *Science*, 1990, **247**, 1461-1465.

9. Abdel-Meguid, S.S., Shieh, H.-S., Smith, W.W., Dayringer, H.E., Violand, B.N. and Bentle, L.A. Three-dimensional structure of a genetically engineered variant of porcine growth hormone. *Proc. Natl. Acad. Sci. USA*, 1987, **84**, 6434-6437.

CATALYTIC ANTIBODIES - CATALYSTS FROM SCRATCH

COLIN J. SUCKLING
Department of Pure and Applied Chemistry, University of Strathclyde,
295, Cathedral Street, Glasgow G1 1XL

INTRODUCTION

It is now well accepted that enzymes can play a valuable role in synthesis on both an industrial and laboratory scale and attention has turned to the production of specific protein-based catalysts at will as well as to optimising the conditions for the use of proteins in organic synthesis. Both problems are under investigation at Strathclyde. The latter is being led by Dr. Peter Halling and this presentation will outline some of the approaches that have been taken to generate new protein-based catalysts suitable for synthetic transformations. We have investigated two approaches, namely *chemically modified enzymes* and *catalytic antibodies*; results from both studies will be described. In both cases, we have chosen to emphasise the generation of new chiral centres in the products of the reactions catalysed and the examples can be considered as prototypes for their respective reactions. Catalytic antibodies, or abzymes, have attracted an especially lively coverage because they offer the potential to produce catalysts for any reaction provided that a reasonable mechanism can be inferred for the reaction. It has been said that it is easier to inhibit an enzyme than to improve one yet paradoxically, the very concepts that have been so successful in the design of enzyme inhibitors also contribute to the field of abzymes. Both build upon the concept of a transition state stabilisation as the basis of enzymic catalysis. If a protein can provide a complementary binding site that lowers the energy of the transition state of a reaction, catalysis will result. The corollary to this is that enzymes bind the transition state strongly. Thus a transition state analogue will bind tightly to an enzyme and inhibit it; conversely, a transition state analogue will also provide the hapten to which an antibody with catalytic activity can be raised. In both cases, the complementarity between protein and ligand leads to the effects observed. Whether one takes an antibody or a normal enzyme, there exists also the possibility of making chemical modifications to the protein to alter its catalytic properties. There are opportunities for novel catalysts using chemically modified proteins although the scope is more limited than for catalytic antibodies.

CHEMICALLY MODIFIED ENZYMES

The concept of introducing a new catalytic function into the active site of an enzyme was rediscovered about 5 years ago by Kaiser [1] who showed how papain could be modified through the incorporation of flavins attached to the active site cysteine. Apart from one reaction, the hydroxylation of aniline, no reactions of synthetic significance were described. We began by considering the possibility of converting papain into an enzyme capable of catalysing acetoin and related condensations (Dr. S.O. Onyiriuka) through the incorporation of a thiazolium salt at the active site. The mechanism of condensation reactions mediated by thiazolium salts is conveniently autocatalytic. By treating papain with n-bromomethyl-1-methylthiazolium bromide, we were able to obtain a modified protein that catalysed the decarboxylation of pyruvate as determined by the coupled enzymic assay of pyruvate (with NADH and lactate dehydrogenase) and acetaldehyde (with NADH and alcohol dehydrogenase) (Figure 1). However, we were unable to demonstrate catalysis of any acetoin-type condensation using a variety of substrates.

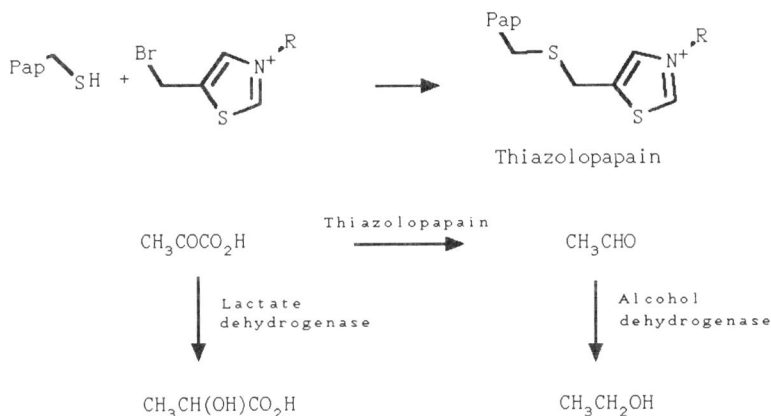

Figure 1.

Such experiments as the above are inevitably speculative and, in order to systematise the investigation, we entered a colloraboration with Dr. Claudine Pascard of ICSN, CNRS, Gif-sur-Yvette, France, supported by the EC. The aims were to explore the incorporation of catalytic groups into papain and subtilisin by varying the chemical structures of the modifying agents, by X-ray crystallography, and by modelling of the modified active sites. So far, we have shown that there is a chemical limitation on the nature of groups that can be incorporated into papain (Figure 2). Thinking that the lack of catalysis of anything other than decarboxylation was due to too close an attachment of the thiazolium salt to papain, we prepared a series of n-bromoethylthiazolium salts. None of these compounds, however, was sufficiently reactive to alkylate papain. Attention has therefore switched for the time being to 3-(2-bromoacetyl)pyridinium salts which are readily available and smoothly alkylate papain according to the kinetics of irreversible enzyme inhibition. Samples of these modified proteins are currently being purified for crystallisation (Dr. Renate Alijah) and their catalytic properties are under investigation (Mr. Zhu Limin). When these results are combined with those from the crystallographic and modelling experiments, we should be able to evaluate rationally the potential for new catalysts based upon chemically modified enzymes.

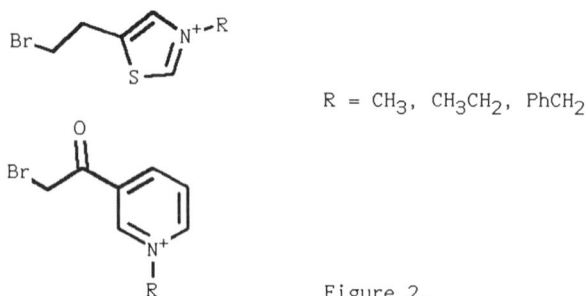

R = CH$_3$, CH$_3$CH$_2$, PhCH$_2$

Figure 2.

CATALYTIC ANTIBODIES

Chemically modified enzymes are limited largely to the insertion of coenzymes or their analogues, and hence the range of reactions that they can catalyse is restricted. The exciting and topical technology of catalytic antibodies offers in principle a method for generating a catalyst for any reaction for which an analogue of a molecule at a critical point on the reaction coordinate can be designed [2]. The most advantageous application of such technology would be in reactions for which there are no readily available enzymes. One such class of reactions concerns electrocyclic reactions and we have studied an example of the Diels Alder reaction. A pair of good reactants has been chosen (Figure 3), the aim of the study being primarily to demonstrate the potential of catalytic antibodies to generate polyfunctional intermediates with many chiral centres. The Diels Alder reaction sets up reliably the relative stereochemistry of substituents in a polycyclic system. Thus the function of the abzyme is to catalyse a reaction leading to only one of the possible enantiomers of the product. In order to approach this goal, the hapten used to raise the antibodies was the N-(4-carboxybutyl) analogue of the product and was synthesised by Dr. Catriona Tedford. The antibodies were raised by Dr. Laura Bence in the laboratories of Professor W.H. Stimson by coupling the hapten to bovine serum albumin and immunising NZB/BALB/C F_1 female mice (8-12 weeks). Clones of antibodies that bound the hapten were identified by conventional ELISA techniques and have been investigated for their ability to catalyse the Diels Alder reaction. Preliminary results have shown that polyclonal sera from immunized mice and cell culture supernatant from cloned monoclonal antibody producing cells have catalytic properties. So far, insufficient protein has been obtained to catalyse reactions on a preparative scale so that the absolute stereochemistry of the products can be determined. This work is in hand.

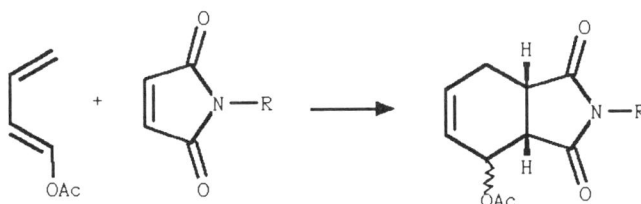

Figure 3.

REFERENCES

1. Kaiser, E.T., Lawrence, D.S., and Rokita, S.E., Annu. Rev. Biochem., 1985, 54, 565-596.

2. Schultz, P.G., Accts. Chem. Res., 1989, 22, 287-294.

ACKNOWLEDGEMENTS

The work described in this paper was supported in part by the European Science Stimulation Programme and Rhone Poulenc Ltd.

ALTERATIONS TO CATALYTIC PROPERTIES OF ENZYMES BY "MEDIA ENGINEERING"

Evgeny N. Vulfson[1], Iqbal Gill[2], Lene Jørgensen[2], Douglas B. Sarney[1],
Igor A. Kozlov[2] and Barry A. Law[1]

[1] AFRC Institute of Food Research, Shinfield, Reading RG2 9AT, UK
[2] University of Reading, Reading, Berkshire RG2 9AT, UK

ABSTRACT

The potential of "media engineering" is illustrated by further examples;
advantages and limitations of this approach are briefly discussed.

The development of enzymatic bioconversions in non-aqueous media is one
of the most exciting recent achievements of biotechnology. Only six years
have passed since the catalytic activity of lipase in dry hydrocarbons
was reported [1] but a number of applications have already been picked up
by industry; for instance, enzyme-catalyzed interesterification of oils,
synthesis of the artificial sweetener aspartame, resolution of racemic
mixtures and polymerization of phenols. This success is due not only to a
solubility factor and/or an equilibrium shift, provided by a drastic
decrease in water activity, but also to the ease of product recovery,
removal of requirements for immobilization and markedly enhanced
operational stability displayed by enzymes in low-water environments.
Another advantage of biotransformations in low-water media is the ability
to alter the catalytic properties of enzymes by changing the composition
of the reaction medium. The feasibility of "media engineering" has been
already demonstrated by a number of examples. Thus, specificity of
peptidases towards certain substrates in organic solvents can be reversed
or drastically shifted, lipases catalyse peptide synthesis, while
subtilisin has been used for the production of sugar esters (see [2] for
a recent review). Even the stereoselectivity of reactions catalyzed by
the same enzyme can be altered by the solvent composition of the reaction
medium [3]. This paper describes several new examples of "media
engineering" and the approach in general is discussed.

RESULTS AND DISCUSSION

Site-specific hydroxylation of aromatic compounds in organic chemistry is notoriously difficult. Available reagents are very unstable and explosive and the product yield is frequently low. The problem is particularly severe when the compounds to be hydroxylated (or their products) are optically active and/or unstable. Alternatively, hydroxylations can be carried out enzymatically by polyphenoloxidase (PPO), but until now it has only limited applicability due to low substrate solubility, drastic substrate inhibition and operational instability of the enzyme. These difficulties may be avoided by using recently introduced detergentless microemulsions as media for biocatalysis [4]. PPO activity has been studied in two types of detergentless microemulsions: hexane-isopropanol-water (Hex/IPA/Wat) and toluene-isopropanol-water (Tol/IPA/Wat); catalytic properties displayed by the enzyme in water and in detergentless reverse micelles has been compared, leading to the following observations:

i) The fast irreversible inactivation of PPO by quinone products could be significantly slowed down in microemulsions with a low water content. As a result the lifetime of the biocatalyst was increased by more than one order of magnitude.

ii) Severe substrate inhibition observed in water could be practically abolished in detergentless reverse micelles. An optimal substrate concentration as well as advantageous K_m values could be easily achieved by manipulation of the composition of microemulsions, and apparent values were found to be dependent on both the ratio of solvents used and on the water content of the system (Table 1).

TABLE 1
Kinetic properties of PPO in water and detergentless reverse micelles

Medium composition*	V_{max}, $\mu molmin^{-1}mg^{-1}$	K_m, mM	C_{I50}, mM**
phosphate buffer pH 6.5	60	0.2	5
Hex/IPA/Wat (45.9/47.8/6.3)	15	3.7	50
Hex/IPA/Wat (47.2/50.8/2.0)	4.0	6.4	80
Tol/IPA/Wat (55.1/42.1/2.8)	3.0	24	>100

* The composition of the reaction medium is expressed in volume %.
** Concentration of 4-methyl catechol (substrate) showing 50% inhibition

Another reaction of great practical interest is enzymatic peptide synthesis. Traditionally, N-protected amino acids have been used as acyl donors due to a relatively low reactivity displayed by endopeptidases towards substrates containing free amino groups, and to the additional shift of equilibrium achieved. However, an elimination of positive charge on an α-amino group by a low-water medium has led to a change in enzyme specificity towards N-unprotected amino acids (Table 2) so that much cheaper substrates can be used for proteinase catalyzed reactions in organic solvents. The value of this methodology for organic synthesis was proved in the preparation of a range of L-dopa esters, currently undergoing clinical trials for the treatment of Parkinson's disease, and we are now studying its applicability to peptide synthesis.

TABLE 2

Specificity of α-chymotrypsin in water and IPA, containing 2% of water

Substrate	Hydrolysis*			Transesterification		
	K_{cat}, s^{-1}	K_m, M	$K_{cat}/K_m, M^{-1}s^{-1}$	K_{cat}, s^{-1}	K_m, M	$K_{cat}/K_m, M^{-1}s^{-1}$
N-Ac-Tyr-OEt	185	1.03×10^{-3}	1.8×10^5	2.19	0.187	11.71
Tyr-OEt	21	6.60×10^{-3}	3.2×10^3	2.71	0.314	8.63

* The reaction was carried out at pH 7.0 at 37°C.

One of the apparent limitations of "media engineering" is the strong preference often displayed by enzymes for a particular organic solvent, as well as a demand for a certain amount of water. This sometimes results in technological and legal restrictions, especially severe in the food industry. This limitation can be eased by changing the physical state of the biocatalyst preparation or the reaction conditions used. For example, the support chosen for immobilization of alcohol dehydrogenase signficantly alters the dependence of its activity on the hydrophobicity of the organic medium. Similar alterations can be achieved by exposure of the enzyme suspension to ultrasound. We have observed a substantial enhancement of subtilisin-catalyzed interesterification under sonication conditions in a wide range of organic solvents, including those regarded as very poor media for biotransformations. Ultrasound treatment of enzyme powder suspended in organic solvents led to the development of maximal activation by added water and/or to suppression of inhibition due to build-up of excess water around enzyme particles (Table 3).

TABLE 3

Enhancement of subtilisin (1.5mg/ml) catalyzed transesterification
of 10mM N-Ac-Phe-OEt in octanol by ultrasound irradiation

Reaction conditions		N-Ac-Phe octyl ester yield, mM			
		0% water	0.5% water	1.0% water	1.5% water
Shaking 200rpm	60min	1.3	1.0	0.2	<0.1
Continuous sonication*	60min	1.6	4.0	3.7	2.0
Sonication 20min/shaking	40min	1.4	3.2	3.0	0.9

* Sonication was carried out at 6 microns peak to peak.

However there are other limitations which probably cannot be
overcome by "media engineering" alone. For instance, several esterases
have been shown to catalyse the formation of sugar-fatty acid esters
[5,6], but enzymatic one-step synthesis of sucrose esters, widely used as
emulsifiers, has not been developed so far. The problem is that lipases
readily catalyze a transfer of fatty acids to a wide range of mono-
saccharides, but not to disaccharides, while subtilisin, which accepts
disaccharides, fails to transfer fatty acids containing more than ten
carbon atoms. Only appropriate changes in enzyme structure through
protein engineering could make this reaction feasible, and a
complementary approach is certainly needed to bring such reactions to
industry.

REFERENCES

1. Zaks, A. and Klibanov, A.M., Enzymatic catalysis in organic media at
   100°C. Science, 1983, 244, 1249-1251.

2. Klibanov, A.M., Enzymatyic catalysis in anhydrous organic solvents:
   kinetics, protein engineering and medium optimisation. Trends
   Biochem. Sci. 1989, 17, 141-144.

3. Kitaguchi, M., Fitzpatrick, P.A., Huber, J.E. and Klibanov, A.M.
   Enzymatic resolution of racemic mixtures: crucial role of the
   solvent. J. Am. Chem. Soc., 1989, 111, 3094-3095

4. Khmelnitsky, Y.L., Hilhorst, R. and Veeger, C., Detergentless
   microemulsions as media for enzymatic reactions. Eur. J. Biochem.,
   1988, 176, 265-271.

5. Riva, S., Chopineau, J., Kieboom, A.P.G. and Klibanov, A.M.,
   Protease-catalyzed regioselective esterification of sugars in
   anhydrous dimethylformamide. J. Am. Chem. Soc., 1988, 110, 584-589.

6. Carrea, G., Riva, S., Secundo, F. and Danieli, B., Enzymatic
   synthesis of various 1'-O-sucrose and 1-O-fructose esters. J. Chem.
   Soc. Perkin Trans., 1989, 1, 1057-1061.

INDOLE GLUCOSINOLATES AND PHYTOALEXINS IN CRUCIFEROUS CROPS

A. BRYAN HANLEY, KEITH R. PARSLEY, JENNY A. LEWIS AND G. ROGER FENWICK
AFRC Institute of Food Research,
Colney Lane, Norwich, Norfolk NR4 7UA, UK

ABSTRACT: A novel breakdown product from indole glucosinolates has been identified and the relationship of this compound to indole phytoalexins investigated.

Indole glucosinolates are a group of tryptophan-derived thioglucosides which occur throughout the Cruciferae[1]. Tissue disruption brings the thioglucosides into contact with a co-occurring thioglucosidase (myrosinase; thioglucoside glucohydrolase E.C. 3.2.3.1) which hydrolyses the C,S, thioglucoside linkage generating glucose and an unstable aglucone. By analogy with glucosinolates which do not possess an indole side chain spontaneous Lossen-type rearrangement is proposed to give indolyl-3-methylisothiocyanate which is then hydrolysed furnishing indole-3-carbinol and thiocyanate ion. The intermediacy of the isothiocyanate in the breakdown of indole glucosinolates remains speculative and is based upon analogy rather than direct evidence (e.g. isolation, trapping experiments etc.).

A second group of indolic secondary metabolites have recently been isolated from cruciferous crops which have been subjected to fungal attack or uv irradiation[2]. The parent compound of this group of phytoalexins is a methyldithiocarbamate and can be formally derived from the addition of a thiomethyl moiety to an indolyl-3-methylisothiocyanate. Incorporation of an intact thiomethyl group has been proposed in the biosynthesis of S-methylcysteine sulphoxide in members of the genus Allium[3] and postulated in Chrysanthemum species[4].

We have attempted to confirm the intermediacy of
indolyl-3-methylisothiocyanates in the enzymic breakdown of indole
glucosinolates and the link with indole phytoalexins by two routes. By
analogy with other indole compounds, we propose that an N-methoxy
substituent will deactivate the isothiocyanate and make it less susceptible
to hydrolysis.

Furthermore, carrying out the hydrolysis in a "low water" system,
where the amount of free water which can participate in the elimination of
thiocyanate ion is limited, will stabilise the intermediate isothiocyanate.

First of all, we have isolated N-methoxyindolyl-3-methylglucosinolate
(neoglucobrassicin) and hydrolysed this compound with myrosinase in a "low
water" system comprising the enzyme and substrate adsorbed onto celite and
then suspended in hexane. Examination by mass spectrometry and subsequent
hydrolysis confirmed the intermediary of N-methoxyindolyl-3-methyl
isothiocyanate in the breakdown of N-methoxyindolyl-3-methylglucosinolate.
Reaction of this relatively unstable intermediate with methanethiol under a
variety of conditions did not afford the methyldithiocarbamate, rather
leading to the elimination of thiocyanate ion and formation of the
corresponding indolylmethyl sulphide (a process analogous to that produced
on aqueous hydrolysis).

We have attempted to synthesise N-methoxyindolyl-3-methyl
isothiocyanate and other breakdown products by a number of routes. While
we have prepared a number of N-methoxy indole compounds including the
3-carbinol, 3-acetonitrile and 3,3'-diindolylmethane (all known indole
glucosinolate breakdown products), we have been unable to synthesise the
isothiocyanate.

Finally, work in this laboratory has demonstrated that mechanical
damage to leaf tissue in cruciferous crops causes a significant increase in
indole glucosinolate levels. Work is presently underway to determine if
there is any correlation between increases in indole glucosinolate levels
and the formation of plant protective indole phytoalexins.

REFERENCES

1. Hanley, A.B., Belton, P.S., Fenwick, G.R. and Janes, N.F., Ring oxygenated indole glucosinolates of Brassica species. Phytochem., 1985, **24**, 598–600.

2. Takasugi, M., Katsui, N. and Shirata, A., Isolation of three novel sulphur-containing phytoalexins from the chinese cabbage Brassica campestris L. ssp. pekinensis (Cruciferae). J. Chem. Soc. Chem. Commun., 1986, 1077–1078.

3. Granroth, B., Biosynthesis and decomposition of cysteine derivatives in onion and other Allium species. Ann. Acad. Sci. Fenn., 1970, **154**, 9–84.

4. Ettlinger, M. and Kjaer, A., Sulfur compounds in plants in Recent Advances in Phytochemistry Vol. 1, ed. T.J. Mabry, R.E. Alston and V.C. Runeckles. North Holland Publishing Co. Ltd., Amsterdam, 1968, pp. 59–144.

THE PRODUCTION OF HETERO-OLIGOSACCHARIDES USING GLYCOSIDASES

R A Rastall, T J Bartlett, M W Adlard and C Bucke

The School of Biological and Health Sciences, The Polytechnic of Central London, 115 New Cavendish Street, London W1M 8JS, UK

At present, there is a great need for supplies of oligosaccharides for research purposes in order to explore their potential as anti-viral agents, as elicitors of various activities in plants and as possible food ingredients. Oligosaccharides can be obtained from a variety of natural sources, but only in small quantities and consequently the costs are high. A potentially inexpensive route for their synthesis is the use of glycosidase enzymes acting "in reverse". The aim of this research program is to determine the ability of readily available glycosidases to synthesise hetero-oligosaccharides. The initial objective is to determine the specificities of the enzymes for substrate structure and size. Much of the information obtained has come from studies using the α-mannosidase from Jack Bean. It has long been known that this enzyme will catalyse transfer and reverse hydrolysis (condensation) reactions [1] , but until recently, no systematic study of the extent to which this enzyme could be used in oligosaccharide synthesis has been made.

Jack Bean α-mannosidase can catalyse the synthesis of oligosaccharides by means of two related reactions, glycosyl transfer and condensation. In the former reaction, the enzyme hydrolyses a reactive substrate, eg: p-Nitrophenyl α-mannoside to produce a glycosyl- enzyme intermediate. The mannose moiety of this intermediate can then be transferred to a suitable acceptor. For convenience in these studies, we have used a range of p-Nitrophenyl glycosides as acceptors. The latter method of oligosaccharide synthesis occurs in the presence of high concentrations of mannose (75 - 85% w/w). The enzyme catalyses the condensation of monosaccharide residues to form di- and trisaccharide products [2] . We have also found that the enzyme is capable of condensing mannose to a range of other sugars when incubated at elevated temperature in mixtures of mannose and acceptor sugar over a wide range of sugar concentrations (45 - 75% w/w total sugar).

The reaction conditions for oligosaccharide synthesis by glycosyl transfer to an acceptor glycoside were optimised using p-Nitrophenyl-β-Galactoside as acceptor, and the principle products formed were characterised by Proton-NMR spectroscopy. These were found to be Man-α 1→6-Gal-β-pNO$_2$ and Man-α-1→4-Gal-α-pNO$_2$. However, product yields were low and only a narrow range of glycosides would act as acceptors, so the "kinetic" approach to oligosaccharide synthesis proved to be of limited value.

The use of the "equilibrium" approach to produce oligosaccharides using α-mannosidase has proved more fruitful. When supplied with mannose alone, the enzyme catalyses the synthesis of α-1,1, α-1, 2, α-1, 3 and α-1,6 mannobioses and higher oligosaccharides. This confirms the results of Johansson et al [2] that no α-1, 4 mannobiose accumulates. Other sugars fell into three categories as acceptors for mannose residues in condensation reactions:

1) Hetero-oligosaccharides formed at both 45% (w/w) and 75% (w/w) total sugar concentration.

2) Hetero-oligosaccharides formed at 45% (w/w) but oligosaccharide synthesis was inhibited at 75% (w/w).

3) No hetero-oligosaccharides formed at either concentration. Mannose condensation not inhibited.

The structures of the hetero-oligosaccharide products of these condensation reactions are currently under investigation. The largest acceptor identi-fied so far is maltotriose. Further exploration of size specificity awaits the availability of larger oligosaccharides at affordable prices.

Other enzymes, most notably, α-glucosidase (yeast), β-glucosidase (Almond), laminarinase (Penicillium) and glucoamylase (Aspergillus niger), have also been studied. Of these, only two (β-glucosidase and glucoamylase) catalyse condensation reactions, and none show evidence of glycosyl transfer. β-glucosidase reversal has been demonstrated previously [3], whilst the reversal of glucoamylase has been extensively studied (for example, see ref 4), although no attempt was made in these studies to co-condense glucose with other sugars. We have found that the β-glucosidase from Almond cannot catalyse co-condensation, and that the glucoamylase from A. niger can co-condense glucose with a few sugars (fucose, xylose, arabinose, fructose, tagatose, sucrose and trehalose), and the structures of these co-condensation products are currently being analysed. The kinetics of reversal of β-glucosidase are also being investigated. In addition, the synthetic potential of A niger β-xylosidase is being assessed and so far, we have found that it is capable only of forming xylose oligosaccharides by condensation of xylose.

Reaction mixtures comprising up to 75% (w/w) sugars are thick syrups which are not amenable to scale-up. To attempt to improve yields of oligo-saccharides in a more practical system, use of 2-phase aqueous systems [5] have been introduced. Work has focussed on the use of the common phase forming polymers Polyethylene glycol (PEG) and dextran. Using the jack bean α-mannosidase, we have found that the bulk of the enzyme partitions into the lower, dextran rich phase. By manipulating the PEG and dextran concentrations, it was possible to obtain a system whereby 95% of the enzyme partitioned into a lower phase comprising 15% of the total volume of the system, whilst p-Nitrophenyl glycosides and p-Nitrophenyl oligosaccharide glycosides produced by enzymatic transfer partitioned more evenly throughout the system. In comparison with transfer reactions in buffer alone, the total yield of product in 2-phase systems was increased by a factor of 10, presumably because, as the bulk of the product was partitioned into the phase with a low level of enzyme, little hydrolysis of the product occurs. Since the bulk of the enzyme remains in the lower phase after separation, this leads to the possibility of continuous synthesis of glycosides and oligosaccharides by exchanging the upper phase containing product with fresh upper phase containing donor and acceptor. The use of 2-phase aqueous systems as media for enzyme catalysed reactions is in its infancy and, given time and funding, much greater sophistication can be introduced.

REFERENCES

1) Li, Yu-Teh (1967) J Biol Chem 242(23), 5474 - 5480.

2) Johansson, E et al (1989) Enz Microb Technol 11, 347 - 352.

3) Dedonder, R A (1961) Ann Rev Biochem 30, 347 - 382.

4) Nikolor, Z L et al (1989) Biotechnol Bioeng 34, 694 - 704.

5) Walter, H Brooks, DE and Fisher, D (eds) "Partition in aqueous 2-phase systems, theory, methods, uses and applications to biotechnology" (Acad. Pr. 1985).

RHIZOPUS NIGRICANS IN THE SYNTHESIS OF A RIA HAPTEN

Alexander KASAL, Jenny POKORNÁ, Jaroslav ZAJÍČEK
Institute of Organic Chemistry and Biochemistry.
Czech.Acad.Sci., 166 10 Prague, Czechoslovakia.

ABSTRACT

Feeding 6β-hydroxy derivative (II) to Rhizopus nigricans affor-
ded 11α-hydroxy derivative (III), its chemical transformation
was complicated by lack of selectivity of the reactions involved. However, feeding 6β-methoxy compound (VIII) yielded the
corresponding 11α-alcohol (IX) with only a single secondary hydroxy group available for acylation, thus subsequent chemical
transformation into the hapten (VII, 3β,11α,17α-trihydroxy-
pregn-5-en-20-one 11 hemisuccinate) proceeded quite readily.

INTRODUCTION

Various cases of hormonal disorders may now be easily identified by quantitative analysis of individual intermediates of
the biosynthesis of steroid hormones. Until now a diagnostic
tool has been lacking in clinical endocrinological laboratories which would make radioimmunoassay of 17α-hydroxypregneno-
lone (3β,17α-dihydroxypregn-5-en-20-one, I) possible. The idea
of linking a steroid molecule to bovine serum albumine via the
11α-hemisuccinyloxy bridge seemed to be the most promising as
this spacer has been proved several times to be the ideal one
when sensitivity to all physiologically important sites of the
molecule was required.

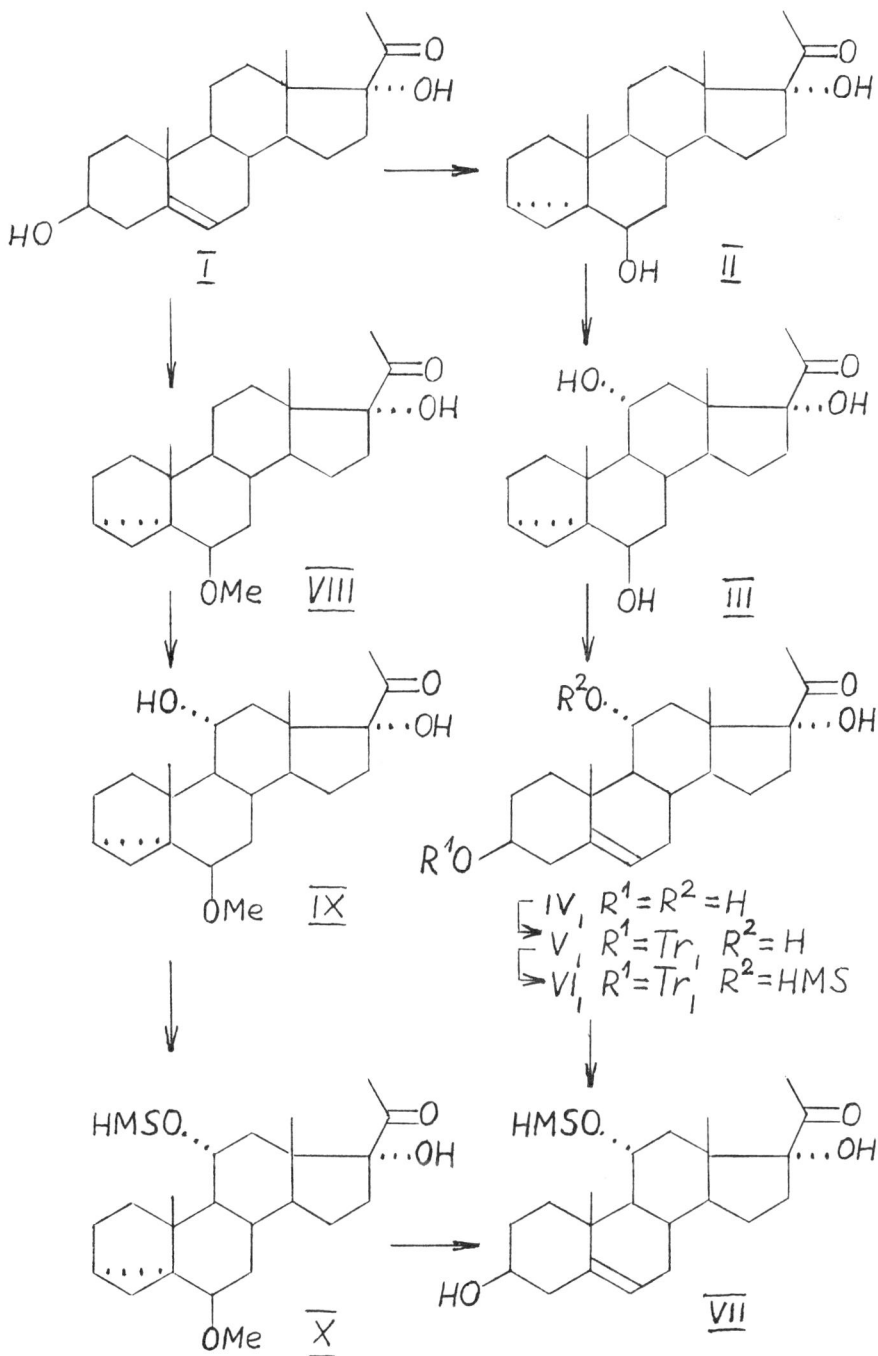

$\begin{array}{l} IV, \quad R^1 = R^2 = H \\ V, \quad R^1 = Tr, \quad R^2 = H \\ VI, \quad R^1 = Tr, \quad R^2 = HMS \end{array}$

RESULTS

It has been known for a long time [1] that the major product of
the action of Rhizopus nigricans on 5-unsaturated steroids is
the allylic hydroxylation in the position 7. If however, the
starting material (I) is first transformed to its synthetic
equivalent, a 6β-hydroxy-3α,5-cyclo-5α-steroid (II), then
expedient 11α-hydroxylation is achieved [2]. Following the procedure described by Protiva, Schwarz and Syhora [3], we obtained
the desired triol (III) in 55% yield. The subsequent transformation into the target molecule (VII) was not straightforward,
it required the preliminary protection of the more reactive 3β-
hydroxy group e.g. by tritylation to (V) which, however, proceeded either incompletely or under more forcing conditions
with the loss of selectivity (both secondary groups in positions 3 and 11 were tritylated simultaneously). The yield of
the hemisuccinate (VII) was therefore only negligible.

An alternative route was then developed which utilized the
6β-methoxy-3α,5-cyclosteroid (VIII) as the starting material in
the microbiological step. The major product of the hydroxylation was the 6β-methoxy derivative (IX) which had only a single
secondary group in its molecule: its esterification could be
carried out without any loss due to concurring reactions at
other sites of the molecule. Conditions were then worked out
which enabled us to transform the 3α,5-cyclo grouping in the
compound (X) into the 5-unsaturated system without breaking
either the ester grouping in the position 11 or the sensitive
α-ketol system in the side chain of the molecule.

Proper choice of the substrate for the microbiological step
thus made the course of subsequent transformations easier.

REFERENCES

1. Kramli A., Horvath J., Nature, 1949, 163, 219.
2. Wechter W.J., Murray H.C., J.Org.Chem., 1963, 28, 755.
3. Protiva J., Schwarz V., Syhora K., Coll.Czech.Chem.Commun.
   1968, 33, 83.

BIODEGRADATION OF STYRENE BY THERMOPHILIC *BACILLUS* ISOLATES

KAILASH C. SRIVASTAVA
Michigan Biotechnology Institute
P.O. Box 27609
Lansing, Michigan, 48909, U.S.A.

ABSTRACT

Several thermophilic *Bacillus* strains capable of growing on styrene as a sole source of carbon were isolated. Two typical strains, one from municipal sludge (FH90) and other from hot spring (51283) were studied in detail. These strains can grow on styrene, its analogs and some of the intermediates of styrene assimilating pathway. The principal products from styrene were styrene oxide, ß-phenethyl alcohol (ß-PA) and o-hydroxyphenylacetic acid (OHA).

Key words: Thermophilic *Bacillus*, styrene, ß-PA, phenyl acetic acid.

INTRODUCTION

Styrene is an important monomer for manufacturing plastics, synthetic rubbers and resins. Studies on catabolism of styrene in mammals (1,2) have shown its degradation into 2-phenylethanol (ß-phenethylalcohol/ß-PA) and phenylacetic acid [PAA (3)]. Microbial degradation of styrene is, however; not well understood. Omori et al. (4,5) reported that bacteria utilizing aromatic hydrocarbons could not utilize styrene as the sole source of carbon. Shirai and Hisatsuka (6) had difficulty in isolating styrene assimilating microorganisms, although they reported biosynthesis of ß-PA by *Pseudomonas* strain 305-STR-1-4 (6,7). Sielicki et al. (8) demonstrated the formation of ß-PA and PAA by a bacterial consortium growing on styrene as the sole carbon source.

In spite of few studies on styrene catabolism, at least five different pathways are known (Fig. 1). Three of these describe the formation of ß-PA by either direct oxidation (6,7, [Fig. 1 a, 1 b]), or through styrene oxide (6 [Fig. 1 c]). Another mechanism is conversion into PAA via a styrene oligomer (8 [Fig. 1 d]). Lastly, the styrene is catabolized into acetoacetic and fumaric acids through PAA and homogentisic acid by a strain of *Pseudomonas fluorescens* (9 [Fig. 1 e]).

Present study was aimed at exploiting the potential of microbial hydroxylation-oxidation of the side chain of styrene for decontamination of environmental pollutants, and production of speciality chemicals with higher stereospecificity, regio-specificity and yields. Thus far, only *Pseudomonas spp.* have been reported to mineralize styrene and accumulate ß-PA, PAA, and styrene oxide. This presentation describes the biodegradation of styrene and its analogs by thermophilic *Bacillus* sp., especially strain 51283.

MATERIALS AND METHODS

Sediment samples obtained from Yellowstone National Park, U.S.A. and sludge samples obtained locally were enriched at 60°C in a basal medium containing 0.1% styrene and (g/l) NH$_4$Cl, 2; CaCl$_2$, 0.1; MgSO$_4$.7H$_2$0, 1; KH$_2$PO4, 1; NaCl, 2; NaNO$_3$, 1; and Na$_2$ HPO$_4$, 1.5, pH 7.0. Growth of isolates was monitored at 610 nm. Metabolic products were analyzed by GC using a glass column (6 ft x 4 mm I.D., Alltech, IL, U.S.A.) packed with 1% SE 30 on silanized chromosorb G (60-80 mesh). Taxonomical characterization was carried out as outlined by Gordon et al. (10).

RESULTS

Isolation of *Bacillus* strains

Aerobic, gram positive, spore forming rods of varying length growing in the pH range of 6.5 - 7.5 and temperature of 45°C-60°C (opt. 60°C) were isolated from hot spring sediments and municipal sludge. On the basis of biochemical characteristics these were identified as *Bacillus* strains.

55

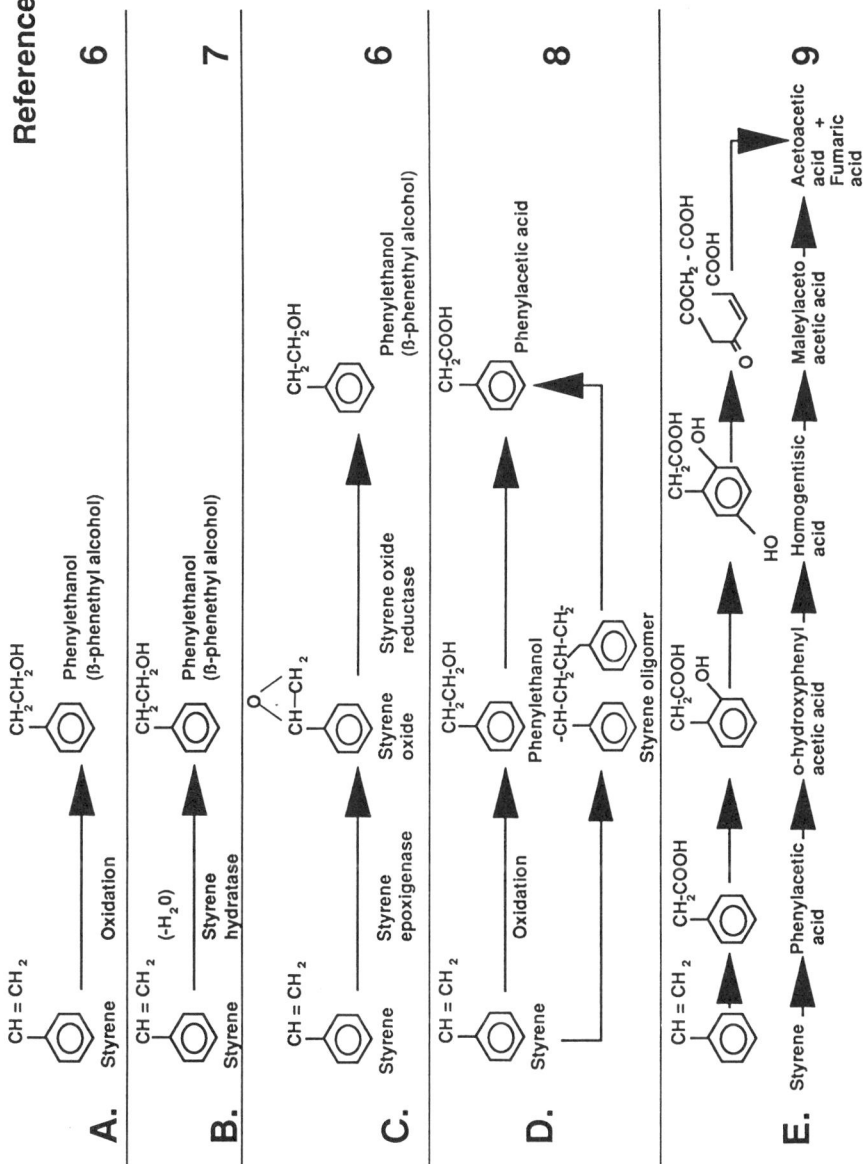

Fig. 1. Suggested Mechanisms for Styrene Degradation

Cultural

Among the amino acids (data not presented), methionine was required for growth. None of the two strains were dependent on alanine, arginine, glycine, leucine, lysine, proline, phenylalanine, serine or valine. While strain 51283 needed (in descending order) cysteine, isoleucine, tryptophane, threonine and tyrosine; strain FH90 was less dependent on threonine and tyrosine but required glutamic acid. Vitamins B_{12}, niacin, biotin, and B_1 were necessary for 51283 but only B_{12}, niacin and biotin for strain FH90. Styrene, styrene oxide or α-methyl styrene at cons. above 2.5% were inhibitory to the growth of both organisms. At 2.5% and above, but at 0.25% and below, production of ß-PA was not observed.

The pH dependence for both, ß-PA production and growth in the range of 5.5-7.0 showed pH 7 to be the best. In the initial lag phase of growth on styrene a white particulate pellicle was observed in the fermenters. This pellicle disappeared after 4-8 hrs of incubation.

Growth on styrene analogs

Both the organisms grew on styrene analogs and intermediates of styrene degradation pathway (Table 1). The major products of growth on styrene and styrene oxide were: ß-PA, PAA and OHA. Styrene oxide was one of the products from both styrene and α-methyl styrene in case of 51283. With FH90, however; styrene oxide was one of the products when grown on α-methyl styrene. PAA and OHA were formed from ß-PA in both the cases.

DISCUSSION

The taxonomical data show the organisms to be thermophilic, neutrophilic *Bacillus* strains. Nevertheless, detailed study of DNA-DNA homology and G+C ratio is warranted to speciate these organisms.

The white pellicle observed in the fermenter was also observed by Sielicki et al. (8). Studies are underway to define this product.

Data obtained from growth on styrene and its analogs (Table 1) are in agreement with those obtained by most of the previous authors (6-9) with respect to the products of styrene catabolism. On this basis, the mechanism of styrene degradation by *Bacillus* strains studied here seem to be the following:

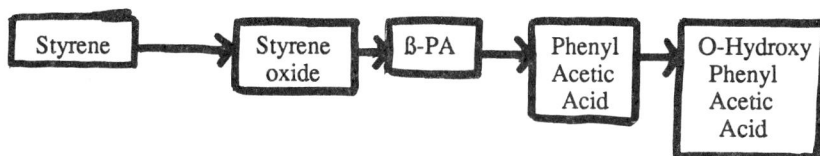

Enzymatic studies are in progress to further elucidate the pathway.

Recently, Bestetti et al. (11) have shown that the products of styrene and α-methyl styrene biodegradation by *Pseudomonas putida* are 2-phenyl-2-propen-1-ol from the oxidation of side chain of α-methyl styrene but 1,2-dihydroxy-3-isopropenyl-3 cyclohexene and 1,2-dihydroxy -3 ethenyl -3 cyclohexene are the products of ring cleavage from α-methyl styrene and styrene, respectively. These are different than those reported thus far (6-9) and for *Bacillus* strains reported here.

TABLE 1

Growth of *Bacillus* strains on styrene, its analogs and
intermediates of styrene metabolism

Substrate	Growth* (Δ O.D.$_{610}$) of strain	
	51283	FH90
Styrene	0.28	1.07
4-Methyl Styrene	0.42	0.20
α-Methyl Styrene	0.73	0.20
Styrene Oxide	1.40	2.83
Phenyl Acetic Acid	0.45	1.13
o-Hydroxy Phenyl Acetic Acid	0.55	0.47
p-Hydroxyphenyl Acetic Acid	0.422	0.27
3,4, Dihydroxyphenyl Acetic Acid	0.82	0.80
Phenethyl Alcohol	0.54	0.77

*All substrates were added at 20 mg/ml conc. The tubes containing 5 ml of medium and appropriate controls were crimp sealed and incubated for 48 hrs at 60°C and 180 rpm.

REFERENCES

1. Pantarotto, C., Farnelli, R., Bidoli, F., Morazzoni, P., Salmona, M., Szcawin Ska, K., Arene oxides in styrene metabolism, a new perspective in styrene toxicity. J. Gen. Microbiol., 1978, **107**, 319-328.

2. Caperos, J.R., Humber, B., Droz, P.O., Exposition au styrene, II Bilan de l'absorption de l'excrétion et du métabolisme sur des sujets humaines. Int. Arch. Occup. Hlth, 1979, **46**, 223-230.

3. Leibman, L.C., Metabolism and toxicity of styrene. Environ. Hlth. Perspect., 1975, **11**, 115-119.

4. Omori, T., Jigami, Y., and Minoda, Y., Microbial oxidation of α-methyl styrene and ß-methyl styrene. Agr. Biol. Chem., 1974, **38**, 409-415.

5. Omori, T., Jigami, Y., and Minoda, Y., Isolation, identification and substrate assimilation specificity of some aromatic hydrocarbon utilizing bacteria. Agric. Biol. Chem., 1975, **39**, 1775-1779.

6. Shirai, K. and Hisatsuka, K., Isolation and identification of styrene assimilating bacteria. Agric. Biol. Chem., 1979, **43**, 1595-1596.

7. Shirai, K. and Hisatsuka, K., Production of ß-Phenethyl alcohol from styrene by *Pseudomonas* 305-STR-1-4. Agric. Biol. Chem., 1979, **43**, 1399-1406.

8. Sielicki, M., Focht., D.D. and Martin, J.P. Microbial transformations of styrene and [^{14}C] styrene in soil and enrichment cultures. Appl. Environ. Microbiol., 1978, **35**, 124-128.

9. Baggi, G., Boga, M.M., Catelani, D., Galli, E., and Treccani, V., Styrene catabolism by a strain of *Pseudomonas fluorescens*. System Appl. Microbiol., 1983, **4**, 141-147.

10. Gordon, R.E., Haynes, W.C., Pang, C.H-N., The Genus *Bacillus*, ARS, USDA, Washington, D.C., 1973, pp. 3-270.

11. Bestetti, G., Galli, E., Benigni, C., Orsini, F. and Pelizzoni, F. Biotransformation of styrenes by a *Pseudomonas putida*. Appl. Microbiol. Biotechnol., 1989, **30**, 252-256.

OPPORTUNITIES FOR BIOTRANSFORMATIONS IN THE FOOD INDUSTRY

SIBEL D. ROLLER, SCOTT J. SWINTON, LEN F.J. WOODS, ROBERT J. HART AND
IAIN C.M. DEA
Biotechnology Unit,
Leatherhead Food Research Association,
Randalls Road, Leatherhead, Surrey KT22 7RY, U.K.

ABSTRACT

Examples of opportunities for biotransformations in the food industry
are given from the biotechnology research programme of the Leatherhead
Food Research Association. Three separate routes for the enzymic
modification of starch were investigated. Firstly, a branching enzyme
was isolated from a food-grade organism and used to increase the degree
of branching of starches in order to reduce the tendency to
retrogradation. Secondly, isoamylase was used to debranch amylopectins
to produce flavour and aroma binding compounds. And, thirdly, the
action of α-amylase was controlled in such a way as to produce
maltodextrins with fat mimetic properties, suitable for incorporation
into low-fat food products.

INTRODUCTION

Encouraged by a growing demand from consumers for foods and food

ingredients that are perceived to be "natural", the food industry is

showing increasing interest in the potential benefits of biotechnology.

At the Biotechnology Unit of the Leatherhead Food Research Association

(LFRA), enzymology and fermentation have been identified as the primary

technologies for achieving desirable biotransformations for the food

industry. These technologies are being used to tailor the molecular

structure of polysaccharides, lipids and proteins. The following paper

highlights some of the opportunities for biotransformations in the food

industry by presenting examples from the LFRA's biotechnology research

programme.

BIOMODIFICATION OF STARCH USING BRANCHING ENZYME

Amylose and amylopectin have a tendency for conformational ordering or

retrogradation - a process responsible for many food deteriorations,

such as the staling of bread. The purpose of this part of the study has

been to introduce additional branch-points into the component molecules

of starch to reduce molecular reassociation and hence extend the
shelf-life of products containing the modified starch.

Classical biochemical techniques, including protamine sulphate and
ammonium sulphate precipitation, dialysis and anion exchange chromatography were used to isolate and purify milligram quantities of branching
enzyme from the food-grade baker's yeast Saccharomyces cerevisiae (1).
Incubation of pure amylose with branching enzyme resulted in a time-
dependent displacement of the absorption maximum of the iodine-amylose
complex from 605 to 560 nm, suggesting the formation of a more branched
polymer from amylose. This observation was not accompanied by an
increase in reducing sugars, indicating that the changes were not due to
contaminating amylolytic activities in the enzyme preparation. Current
research in this area is focused on the scale-up of the branching
reaction to produce enough modified starch for testing in foods.

ENZYMIC DEBRANCHING OF STARCH

The objective of this part of the study was to prepare oligosaccharide
mixtures with aroma and flavour binding properties. Waxy maize starch
and potato amylopectin were debranched using isoamylase from Pseudomonas
spp. A typical reaction mixture consisted of 10% (m/V) solutions of
starch in 0.01M citrate buffer, pH 3.5, incubated with approximately 25
units/ml of isoamylase for 24 h at 45°C. The debranching reaction was
monitored by measuring the release of reducing sugars (2). The enzyme
was inactivated by boiling and the products were freeze-dried. Binding
activity to methyl orange, ethyl orange, metanil yellow, brilliant
yellow, methyl yellow, tropaeolin O and Congo red was measured
spectrophotometrically (3). The binding activity of the debranched
starches was shown to be greater than that of the native starches but
not as great as that of cyclodextrin. Differences in binding capacity
may have been due to the different molecular dimensions of the
hydrophobic cores within the short oligosaccharides and the
cyclodextrin. It is also likely that the debranched starches were a
heterologous population of helices whilst the cyclodextrins are
homogeneous, flat, cyclic rings, resulting in different requirements
for suitable guest molecules for the two compounds.

ENZYMIC MODIFICATION OF STARCH FOR THE PRODUCTION OF FAT MIMETICS

Alpha-amylases can be used for moderate treatments of starch to produce
starch hydrolysis products (SHPs) with dextrose equivalents (DEs) of
less than 10. These SHPs form thermoreversible gels with properties

resembling those of processed fats. The lower calorific value of carbohydrates consequently makes SHPs an attractive alternative to fats in reduced-calorie food products. The objectives of the work carried out at the LFRA were to develop a range of novel ingredients based on the controlled enzymic degradation of starch, to attempt to correlate their chemical, physical and functional properties and to evaluate their effectiveness as fat mimetics in food.

Several 30% solutions of native potato starch were treated with porcine α-amylase in a steam-jacketed mixer for varying periods of time at 60°C. The course of starch hydrolysis was monitored by measuring the release of reducing sugars (2). In this way, it was possible to establish the relationship between processing parameters and degree of hydrolysis. The enzyme reaction was terminated by rapid heating to 95°C followed by freeze-drying of the product. Analysis of final DE values showed that there was a linear relationship between DE and holding time at 60°C during processing. This work established the conditions necessary for the production of SHPs from a specific combination of enzyme and substrate.

Rheological analyses of SHP gels were carried out using a Brabender amylograph, a Stevens Texture Analyser and a Carri-Med Controlled Stress Rheometer (Table 1). The melting point measurements suggested that SHPs with DEs from 3.8 to 6.0 might be suitable as fat replacers for food products that melt in the mouth. However, these melting points were determined by applying a stress force to the gel and then increasing the temperature until the gel yielded to that stress. Typically, a gel will yield at its melting point and this yield temperature will be independent of the stress applied. However, the yield temperatures of the SHP gels were dependent on the applied stress, as shown in Table 2.

TABLE 1
Properties of gels prepared from starch hydrolysis products

SHP	DE	Gel strength* (g/mm)	Peak viscosity** (Brabender Units)	Melting point** (°C)
A	2.2	29.8	970	~60
B	3.8	21.1	350	37
C	5.1	13.0	230	36
D	6.0	13.3	280	38
E	7.1	11.4	240	47

* Measured at 15% (m/V)
** Measured at 10% (m/V)

TABLE 2

Yield temperatures for SHP A (DE=2.2) gel (10% m/V)

Applied stress (N/m^2)	Temperature at which yield occurred (°C)
2.5	70.7
3.0	46.5
5.0	41.6
8.0	37.4
10.0	32.5

The melting points reported in Table 1 give estimates of "softening points" rather than of true melting points. Nevertheless, the rheological analyses alone suggest that changes in the molecular interactions between constituent oligosaccharides had occurred in the different SHPs produced.

One of the SHPs prepared in this study (SHP D, DE=6) was evaluated in low-fat British sausages. Batches of low-fat sausages containing 2.5% SHP and 12.2% fat, and full-fat sausages containing 32% fat were prepared and evaluated by a panel of ten experienced sensory assessors. Three leading brands of commercial low-fat sausages were included in the tasting for comparison. The SHP sausages compared favourably and were, on occasion, preferred to the commercial samples. Further work with different enzymes, starting materials and processing conditions will be required before chemical and physical characteristics predictive of performance in foods can be identified.

REFERENCES

1. Swinton, S.J. and Woods, L.F.J., Extraction and purification of a microbial branching enzyme. Fd. Biotechnol., 1989, 3(2), 197-202.

2. Miller, G.L., Use of dinitrosalicylic acid reagent for determination of reducing sugar. Analyt. Chem., 1959, 31, 426-428.

3. Makela, M.J., Korpela, T.K., Puisto, J. and Laakso, S.V., Nonchromatographic cyclodextrin assays: Evaluation of sensitivity, specificity and conversion mixture applications. J. Agric. Fd. Chem., 1988, 36, 83-88.

STIRRED TANK POWER INPUT DATA FOR THE SCALE-UP OF TWO-LIQUID PHASE BIOTRANSFORMATIONS

J. M. WOODLEY
SERC Centre for Biochemical Engineering,
Department of Chemical and Biochemical Engineering,
University College London,
Torrington Place,
London WC1E 7JE, UK

ABSTRACT

The scale-up of two-liquid phase biotransformations will necessitate the measurement of appropriate laboratory engineering data. In this paper the scale-up of such reactions in stirred tank reactors, using maintenance of power input per unit volume, is discussed.

INTRODUCTION

Biological catalysts are now being employed to carry out a large range of organic reactions and among the most interesting and industrially useful are those involving poorly water-soluble organic compounds [1]. However, many industrially useful reactions involve water-soluble as well as poorly water-soluble reaction components [2,3] and for these cases the use of an heterogeneous liquid-liquid reaction medium is particularly attractive [4]. In parallel with the many reported scientific studies of these systems it is important that appropriate biochemical engineering is examined to assist in the scale-up of these processes. In this paper laboratory power input data and its use in scaling-up two-liquid phase reactions carried out in stirred tanks, is examined.

LABORATORY REACTOR POWER INPUT

Scale-up of stirred tank liquid-liquid mass transfer processes is generally based upon maintaining a constant power input per unit volume, independent of scale [5]. Thus, for those biotranformations occurring in the bulk aqueous phase of a two-liquid phase reaction liquor [6], it is also reasonable to consider scaling-up on a similar basis. In order to assess the feasibility of such a method it is crucial to know what power input is required for a laboratory scale two-liquid phase

TABLE 1

Two-liquid phase enzyme reactor characteristics

Reactor characteristic*	Reactor 1 (R1)	Reactor 2 (R2)
Tank diameter (mm), T	38	50
Impeller diameter (mm)	18	25
Liquor volume (ml)	75	75
Liquor height (mm), H	68	38

*Baffled stirred vessel with 6-blade disc turbine.   System details and reactor operation given previously [7].

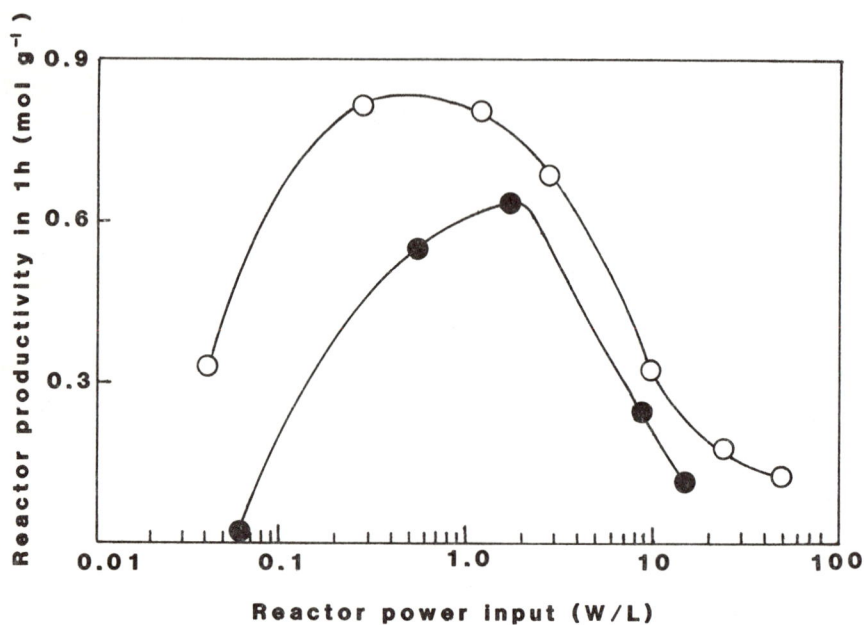

Figure 1.   Menthol produced in 1h, from menthyl acetate hydrolysis by PLE, as a function of power input per unit volume in Reactor 1 (●) and Reactor 2 (○).

biotransformation. In particular, it is important to know whether the power input required for optimum use of catalyst and mass transfer on a laboratory scale is achievable on an industrial scale. Here, this is illustrated with the hydrolysis of menthyl acetate to menthol by pig liver esterase, PLE, which has been reported previously [7]. The reaction was carried out in two different stirred tank reactors, the characteristics of which are given in Table 1.

Organic and aqueous phase properties (estimated, where appropriate [8]) were used to compute effective dispersion properties for density and viscosity, determined using the equations proposed by Mersmann and Grossman [9]. Using these effective dispersion properties the impeller Reynolds number in each tank, as a function of agitator speed, N, was calculated. Above Reynolds numbers of 10,000 (occurring at N = 3600 and 1800 rpm for R1 and R2, respectively) the system may be considered turbulent, with a constant power number of 5. Systems with Reynolds numbers between 10 and 10,000 are in the transition regime with power numbers between 4 and 5. Beneath a Reynolds number of 10 (occurring at N = 35 and 18 rpm for R1 and R2, respectively) the system is in the laminar regime with power numbers between 5 and 50. The power number is read from a chart of power number as a function of Reynolds number for the particular vessel and impeller geometry [10]. The required reactor power input may then be calculated from the power number. For the menthyl acetate hydrolysis by PLE at an aqueous phase enzyme concentration of 0.1 g/L, optimum productivity occurs at power inputs of 1.5 and 0.3 W/L for R1 and R2, respectively, Figure 1. Figure 1 also shows that higher productivities are attained in R2 than R1, regardless of the power input to the two reactors. The two reactors differ geometrically only in aspect ratio (H/T = 1.79 and 0.76 for R1 and R2, respectively). This emphasises the need to maintain complete geometric similarity in addition to maintaining phase ratio, catalyst concentration and power per unit volume constant, during reactor scale-up.

DISCUSSION

Power inputs up to 4 W/L are used in many large scale processes (e.g. 25,000 L) and hence the estimated requirement for the menthyl acetate hydrolysis by PLE is not problematic for scale-up. An increase in enzyme concentration will necessitate a concomitant increase in mass transfer rate and hence power input. At aqueous phase enzyme concentrations as high as 5 g/L power inputs up to 75 W/L (15 W/L for R2) will be required. This is practical up to scales of about 1500 L (6500 L for R2), which may be adequate for the production of many speciality, high value added, products.

Microbially catalysed reactions have added complications. For example, at high cell concentrations the aqueous phase may exhibit non-Newtonian rheological properties, leading to high viscosities, low Reynolds numbers and high power requirements. The hydrolysis of menthyl acetate has also been carried out using *B. subtilis* as a catalyst [7]. Here the aqueous phase was pseudoplastic and had apparent viscosities up to 300 cP at aqueous phase cell concentrations of 100 g/L. Nevertheless, optimum productivity occurred at a power requirement of 18 W/L, scaleable to about 5000 L.

ACKNOWLEDGEMENT

The author is grateful to the Science and Engineering Research Council for support of this work.

REFERENCES

1. Laane, C., Tramper, J. and Lilly, M.D. (Eds), Biocatalysis in Organic Media, Elsevier, Amsterdam, 1987.

2. Ballard, D.G.H., Courtis, A., Shirley, I.M. and Taylor, S.C., A biotech route to polyphenylene. J. Chem. Soc. Chem. Commun., 1983, **634**, 954-5.

3. Hirohara, H., Mitsuda, S. and Hudo, E., Enzymatic preparation of optically active alcohols related to synthetic pyrethroids insecticides. In Biocatalysts in Organic Syntheses, ed. J. Tramper, H.C. van der Plas and P. Linko, Elsevier, Amsterdam, 1985, pp. 119-34.

4. Lilly, M.D. and Woodley, J.M., Biocatalytic reactions involving water-insoluble compounds. In Biocatalysis in Organic Media, ed. C. Laane, J. Tramper and M.D. Lilly, Elsevier, Amsterdam, 1987, pp. 179-92.

5. Carpenter, K.J., Fluid processing in agitated vessels. Chem. Eng. Res. Des., 1986, **64**, 3-10.

6. Woodley, J.M., Two-liquid phase biocatalysis: reactor design. In Biocatalysis, ed. D.A. Abramowicz, Van Nostrand Reinhold, New York, in press.

7. Williams, A.C., Woodley, J.M., Ellis, P.A., Narendranathan, T.J. and Lilly, M.D., A comparison of pig liver esterase and Bacillus subtilis as catalysts for the hydrolysis of menthyl acetate in stirred two-liquid phase reactors. Enzyme Microb. Technol., in press.

8. Souders, M. Viscosity and Chemical consitution. J. Am. Chem. Soc. 1938, **60**, 154-8.

9. Mersmann, A. and Grossman, H., Dispersion of immiscible liquids in agitated vessels., Int. Chem. Engng, 1982, 22, 581-90·

10. Uhl, W.V. and Gray, J.B., Mixing, Academic Press, New York, 1966.

SCALE-UP OF THE PROCESS FOR THE BIOTRANSFORMATION OF NICOTINIC ACID INTO 6-HYDROXYNICOTINIC ACID

FRANS W.J.M.M. HOEKS, HANS-PETER MEYER, DANIEL QUARROZ,
MICHAEL HELWIG AND PAVEL LEHKY

Lonza AG

CH-3930 Visp, Switzerland

ABSTRACT

Oxygen transfer was the rate limiting step in the biotransformation of nicotinic acid into 6-hydroxynicotinic acid. The biotransformation was carried out in reactors with volumes of up to 12'000 liters. Although reaction times differed according to reactor configuration, the 6-hydroxynicotinic acid was of excellent quality and therefore suitable for further chemical syntheses.

INTRODUCTION

As part of the coenzymes NAD and NADP, nicotinamide plays an important role in nature. Substituted nicotinic acid derivatives could be specific inhibitors or modulators of NAD or NADP dependent enzymes and may therefore find applications as pesticides or pharmaceuticals. Nicotinic acid derivatives substituted in position 6 can be produced by chemical synthesis, but the formation and subsequent separation of by-products makes the cost prohibitively high [1, 2].

The first metabolite in the microbial degradation pathway of nicotinic acid is 6-hydroxynicotinic acid [3]. It has been demonstrated that the biotransformation of nicotinic acid into 6-hydroxynicotinic acid by micro-organisms is highly specific and is therefore a good alternative to the chemical synthesis [4].

The starting material for the chemical synthesis of 6-substituted nicotinic acid derivatives (see Scheme 1) is also 6-hydroxynicotinic acid. Two such derivatives of this compound, namely 5,6-dichloronicotinic acid and 2-chloro-5(hydroxymethyl)pyridine, are economically very promising. Both substances are building blocks for the chemical synthesis of insecticides [5, 6, 7]. The expected demand for the year 2000 is several hundreds of metric tons per year.

This poster will discuss the scale-up of the biotransformation process developed by LONZA [4] for the production of 6-hydroxynicotinic acid.

Scheme 1

PRODUCTS DERIVED FROM THE REACTIONS USING 6-HNA AS
THE STARTING MATERIAL

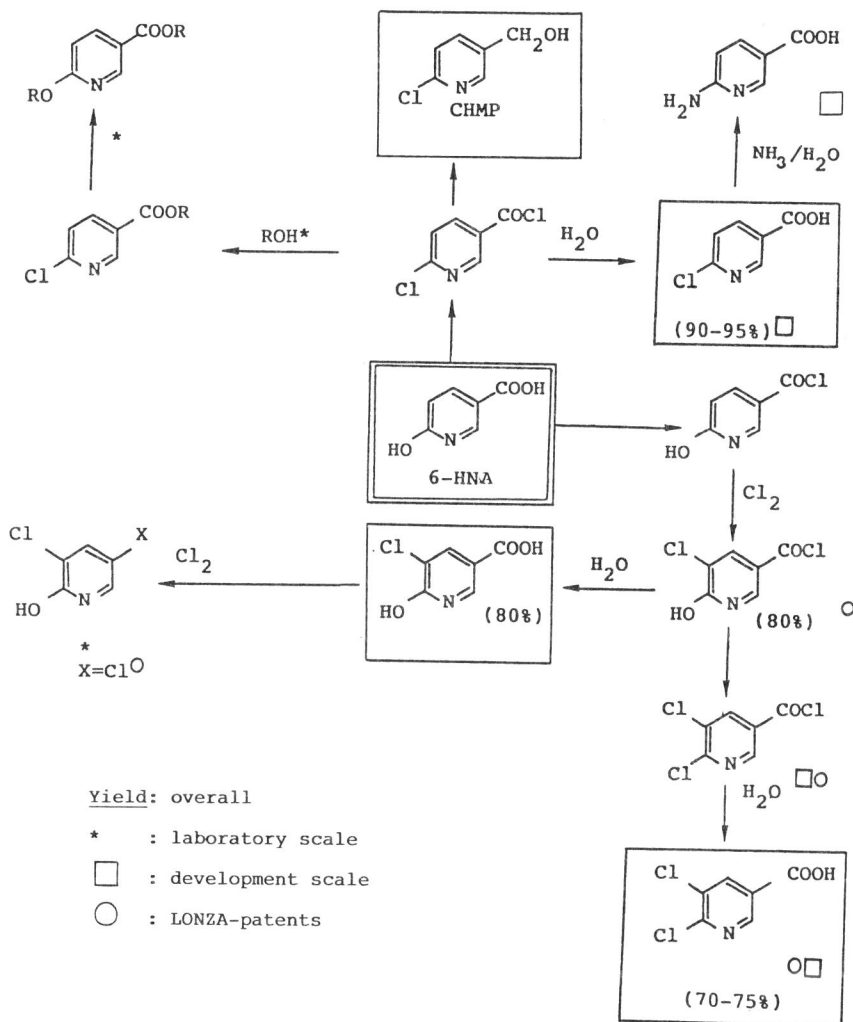

THEORETICAL APPROACH

The first step in the microbial degradation of nicotinic
acid is hydroxylation to 6-hydroxynicotinic acid [8]. The
cofactor for this reaction is the electron carrier NADP. In
aerobic micro-organisms such as Pseudomonas and Bacillus,
the final acceptor of electrons is molecular oxygen [8, 9]
and the (simplified) reaction equation can be written as
follows :

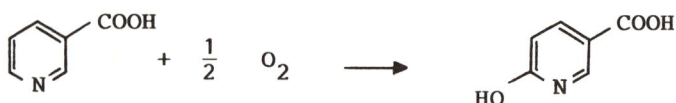

Several biochemical studies have shown that microbial
cells possess a very high hydroxylating activity when
cultivated in the presence of nicotinic acid [9, 10]. When
these data are converted to biochemical engineering terms,
the specific activity might be as high as 20-35 mmol of 6-
hydroxynicotinic acid formed per gram of dry matter of
biomass per hour. According to the reaction equation this
would mean a specific oxygen consumption rate of 10-17.5
mmol of oxygen per gram of dry matter of biomass per hour.

Since a standard bioreactor reaches maximum oxygen
transfer rates of approximately 200 mmol/L/h it is clear
that maximum oxygen transfer rates will be reached at
relatively low biomass concentrations (~10-20 grams of dry
matter per liter) in the hydroxylation reactor. Therefore
the oxygen transfer rate can be identified as the limiting
step in the hydroxylation process.

PRACTICAL RESULTS

The biotransformation of nicotinic acid into 6-hydroxynicotinic acid was carried out in reactors with various volumes in the range between 20 and 12'000 litres, and with several different impeller and aeration configurations. Various reaction times were observed, due to the rate limiting nature of the oxygen transfer process. However, the 6-hydroxynicotinic acid obtained after isolation from the broth was in all cases of excellent quality and could be used directly in further chemical syntheses. The hydroxylation was virtually complete with yields close to 100 %.

REFERENCES

[1]   Briancourt et al., J.Chim.Ther. (1973), 8 (2), 226-232

[2]   D. Quarroz; Patent EU-0084118 (1986)

[3]   E.J. Behrman, R.Y. Stainer: J.Biol.Chem. 228 (1957), 923-946

[4]   P. Lehky, H. Kulla, S. Mischler; Patent EU-0152948 (1985)

[5]   Patent EU-292 822

[6]   Patent EU-302 389

[7]   Patent EU-302 833

[8]   A.L. Hunt, D.E. Hughes, J.M. Lowenstein, Biochem.J. 69, 170-183 (1958)

[9]   J.S. Holcenberg, E.R. Stadtman, J.Biol.Chem. 1969, 244 (5), 1194-1203

[10]   D.E. Hughes, A.L. Hunt, A. Rodgers, J.M. Lowenstein, Proc.Intern.Congr.Biochem., 4th, Vienna 1958 13, 189-93, Pub. 1960

BIOEPOXIDATIONS BY Nocardia OU

J. R. Hunt and H. Dalton
Dept. of Biological Sciences
Warwick University
COVENTRY
West Midlands
CV4 7AL

Ethylene oxide has been used for sterilisation purposes since the 1930's after recognition of its toxicity to microorganisms. This asset illustrates the problem of producing epoxides biologically especially when competing with a chemical process. Perhaps the most important use of epoxides is in the production of chiral pharmaceuticals and agrochemicals. It is at this level that biologically-produced epoxides can compete successfully with chemically synthesised ones since many microorganisms are capable of producing enantiomerically pure products, a phenomenon not generally encountered in chemistry.

An octane-utilizing (OU) strain of *Nocardia* capable of cometabolic alkene epoxidation has been selected by continuous enrichment in chemostat culture. Growth of *Nocardia* OU on glucose showed there was only a low level constitutive expression of epoxidating activity which is significantly induced when grown on n-alkanes. It was shown to grow on a wide range of n-alkanes (propane to hexadecane) expressing a higher specific growth rate on the even chain-length n-alkanes from propane to hexane (inclusive), whereas with the higher n-alkanes a faster growth rate was supported by the odd chain-length hydrocarbons.

The epoxidation activity of *Nocardia* OU was studied most extensively with cells grown in chemostat culture with n-hexane or n-heptane vapour. These cells were shown to epoxidate 1-alkenes from propene to 1-tetradecane, aryl alkenes (styrene, allyl benzene and allyl phenyl ether) and cyclohexane. In the last case, cyclohexene oxide was only a minor reaction product, the other products appearing as a result of an epoxide ring opening reaction to give the allyl alcohol followed by a series of redox reactions.

Specific epoxidation rates were shown to decrease with alkene chain length which may be due to the decrease in alkene solubility.

In common with the other Actinomycetes and related organisms capable of alkene epoxidation, the specific rates of epoxidation by *Nocardia* OU are poor (< 10 nmol 1,2-epoxypropane/min/mg dry weight with heptane grown cells) compared to the rates expressed by the methanotrophs (about 700 nmol 1,2-epoxpropane/min/mg dry weight of *Methylococcus capsulatus* (Bath)). Epoxidation of 1-hexane by hexane-grown cells of *Nocardia* OU compares more favourably with the rates expressed by *Methylococcus capsulatus* (Bath); the latter organism performs only five to six times better than the former at a similar biomass concentration.

The specific epoxidation rate of *Nocardia* OU was found to be inversely proportional to the biomass concentration used in the assay, whilst the overall productivity was generally proportional to the biomass concentration. Assay temperature was also shown to affect the rate of 1-hexene epoxidation, being optimal at around 25°C.

In contrast to the methanotrophs, the Actinomycetes and related organisms are known to produce epoxides with a high degree of enantioselectivity. It is this factor that makes a biotransformation process competitive with its corresponding chemical transformation process. Initial studies with *Nocardia* OU suggest that the epoxidation of styrene by hexane-grown cells is significantly more enantioselective than is the case with *Methylococcus capsulatus* (Bath).

Growth of *Nocardia* OU on substrates other than n-hexane may alter the organism's substrate specificity and this area is currently under study.

INVESTIGATION OF MONOHYDROXYLATION REACTIONS IN BENZ[+] STRAINS OF *Pseudomonas putida*

S. E. Jones and H. Dalton
Department of Biological Sciences
Warwick University
COVENTRY
West Midlands
CV4 7AL

Benzene and toluene dioxygenases are multi-component enzyme systems comprising a flavoprotein component, ferredoxin component and an oxygenase component, and have been isolated and purified from *Pseudomonas putida* (1-4). These dioxygenases catalyse the initial step in the degradation of aromatic substrates, by incorporation of both atoms of molecular oxygen into the substrate, yielding the corresponding *cis*-glycol (4-6) as illustrated in Figure 1.

FIGURE 1 :   THE INITIAL STEPS IN THE OXIDATIVE DEGRADATION OF BENZENE IN *P. putida*.

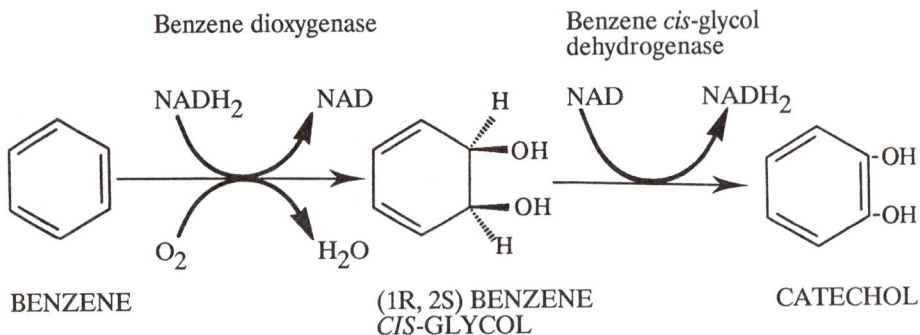

BENZENE   (1R, 2S) BENZENE   CATECHOL
                           CIS-GLYCOL

Recently, monooxygenation type reactions catalysed by strains of *P. putida* expressing dioxygenase activity have been reported (7, 8). Wackett *et al*. (7) have reported that a strain of *P. putida* expressing toluene dioxygenase is able to metabolise indene yielding not only the expected *cis*-diol but also l-indenol and 1-indanone, whilst the metabolism of indan yields l-indanol and l-indanone.

The formation of l-indanone in indene metabolism was found to result from non-enzymic isomerisation of l-indenol. However, in the case of indan metabolism, l-indanone was found to be derived from l-indenol by means of a toluene inducible dehydrogenase reaction. Studies with $^{18}O_2$ and $H_2^{18}O$ indicated that oxygen atoms in l-indenol and *cis*-1,2-indandiol were derived from molecular oxygen whilst 70% of oxygen in l-indanol was derived from water. However, lack of catalytic activity in the absence of oxygen, the requirement for NADH to catalyse monooxygenation and the stimulation of the monooxygenation and dioxygenation reactions by ferrous ions indicates that oxygen activation by toluene dioxygenase is a prerequisite for catalysis. Monooxygenation products were also obtained when purified toluene dioxygenase was used indicating the dioxygenase itself is able to catalyse monooxygenation type reactions as well as dioxygenation.

With this in mind a variety of potential substrates have been screened with strains of *Pseudomonas putida* to investigate the range of mono- and dioxygenation reactions catalysed. Further studies will then be made with the purified enzyme system to obtain a more detailed understanding of the mechanism of the monooxygenation reactions.

1, 4-dihydronaphthalene, cyclohexene, indene and benzylmethylsulphide, for example, all yield a mixture of mono and dioxygenated products upon incubation with strains of *Pseudomonas putida* expressing dioxygenase activity. Relative yields and optical purities of products are, in many cases, high. For example, the biotransformation of 1,4-dihydronaphthalene by *P. putida* UV4 yields two monooxygenated and one dioxygenated product as indicated below.

FIGURE 2 :   RELATIVE YIELDS AND e.e.'s OF PRODUCTS FORMED IN
              BIOTRANSFORMATIONS OF 1,4-DIHYDRONAPHTHALENE.

Incubation Time	Starting Material	β-Hydrate	Pseudohydrate	CIS-Dihydrodiol
30 min	48%	4%	48%	NIL
22 hrs	NIL	25%	63%	12%
e.e.		<5%	>99%	>99%

The ability of the organism to catalyse monohydroxylation and dioxygenation reactions is dependent upon the substrate. Benzene, toluene and naphthalene give rise only to their corresponding *cis*-dihydrodiols and not phenol, methylphenol or naphthol. Aromatic substrates such as quinoxaline, quinoline, isoquinoline and quinazoline however yield both *cis*-dihydrodiols and monohydroxylated products i.e. phenols (8). Monohydroxylations also occur in non-aromatic systems (e.g. 1,4-DHN etc.). In some cases only monooxygenation products have been detected, for example with various cyclic disulphides.

i.e.

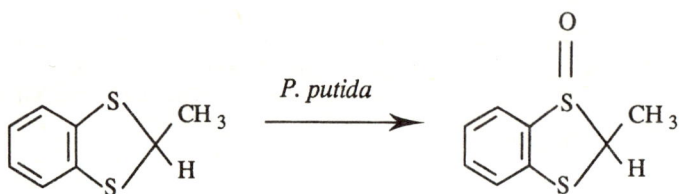

In such cases it is presumed that the substrate is unsuitable for dioxygenation due to its steric or electronic properties.

The products of such reactions may also be influenced to some extent by use of a particular strain of *Pseudomonas putida*. This is indicated below using the example of the biotransformation of benzylmethylsulphide.

THE BIOTRANSFORMATION OF BENZYLMETHYLSULPHIDE BY VARIOUS STRAINS OF *P. putida*.

STRAIN	CARBON SOURCE	INCUBATION TIME (HRS)	CH₂SMe	CH₂SMe ⟨OH OH⟩	O CH₂SMe	COOH
UV4	Succinate	2	-	70%	7.5%	22.5%
	Succinate	22	-	83%	8.5%	8.5%
UV4	Succinate	2	70.5%	1%	1%	27.5%
	+ Toluene	22	-	78%	9%	13.0%
NG1	Succinate	2	-	70%	30%	-
	+ Toluene	22	-	37%	63%	-
ML2	Benzene	2	-	98%	2%	-
		22	-	100%	-	-
11767	Succinate	2	-	-	29%	71%
	+ Toluene	22	-	-	13%	87%
11767	Toluene	2	-	-	47%	53%
		22	-	-	50%	50%

P. putida strains ML2 and 11767 are both wild type organisms yet give very different results when incubated with benzylmethylsulphide, 11767 giving no *cis*-diol whilst ML2 gives almost exclusively *cis*-diol. These two strains may therefore be of use in investigations of the mechanism of monohydroxylation.

Initial experiments on the purification of benzene dioxygenase yielded an active crude cell extract, the dioxygenase activity of which could be measured polarographically. The crude extract also displayed the ability to catalyse the monooxygenation and dioxygenation of indene and the monooxygenation of a cyclic disulphide. However attempts to purify further the enzyme by means of FPLC were accompanied by a loss of activity.

Further attempts will be made to purify the components of the dioxygenase system so that their ability to catalyse monooxygenation reactions may be assessed.

PREPARATION OF CRUDE BENZENE DIOXYGENASE EXTRACT FROM *Pseudomonas putida* ML2.

PURIFICATION STEP	TOTAL VOL. (ml)	TOTAL PROTEIN (mg)	TOTAL ACTIVITY (units)	SPECIFIC ACTIVITY (units)	RECOVERY (%)
CRUDE CELL EXTRACT	98	2734	33817	12.37	100
PROTAMINE SULPHATE SUPERNATANT	97	3055	39656	12.98	117
40-70% (W/V) SATD. $(NH_4)_2SO_4$ FRACTION (Dialysed)	26	1115.7	24919	22.34	73

REFERENCES

1. Axell, B. C. & P. J. Geary, Biochem. J. 146, 173 (1975).
2. Yeh, W. K., D. T. Gibson & Te-Ning, Liu, Biochem & Biophys. Res. Commun., 78, 1 (1977).
3. Subramanian, V., Te-Ning, Liu, W. K. Yeh, C. M. Serdar, L. P. Wackett & D. T. Gibson, J. Biol. Chem. 260, 4 (1985) 2355.
4. Zamanian, M. & J. R. Mason, Biochem. J. 244, 611 (1987).
5. Gibson, D. T., M. Hensley, H. Yoshioka, T. J. Mabry, Biochem. 9, 626 (1970).
6. Gibson, D. T., G. E. Cardine, F. C. Maseles & R. E. Kallio, Biochem. 9, 1631 (1970).
7. Wackett, L. P., L. D. Kwart & D. T. Gibson, Biochem. 27, 1360 (1988).
8. Boyd, D. B., R. Austin, S. McMordie, H. P. Porter, H. Dalton, R. O. Jenkins & O. W. Howarth, J. Chem. Soc. Commun. (1987) 1722-1724.

BIOTRANSFORMATION OF STYRENE

S. HARTMANS, M.J. VAN DER WERF AND J.A.M. DE BONT
Agricultural University, Division of Industrial Microbiology
P.O. 8129, 6700 EV Wageningen, The Netherlands.

ABSTRACT

Using very low concentrations of styrene as sole source of carbon and energy 14 aerobic bacteria and 2 fungi were isolated. In cell extracts of 11 of the bacterial isolates a novel FAD-requiring styrene monooxygenase activity was detected that oxidized styrene to styrene oxide (phenyl oxirane). In one bacterial strain (S5), styrene metabolism was studied in more detail. In addition to styrene monooxygenase, cell extracts from strain S5 contained styrene oxide isomerase and phenylacetaldehyde dehydrogenase activities. Using styrene-grown cells accumulation of substituted epoxystyrenes was possible.

INTRODUCTION

Chiral epoxides are valuable intermediates in the synthesis of a wide range of optically active drugs. The possibilities of direct chemical synthesis of chiral epoxides are however at present still very limited. Enzymatic epoxidation of the carbon-carbon double bond of a wide range of molecules has been demonstrated using bacterial cells containing monooxygenase activities [1]. Most of the work has focussed on methane-, alkene- and alkane-utilizing microorganisms. The enantiomeric purity of the epoxides formed depends on both the substrate and the monooxygenase used [2][3][4].

Styrene oxide has previously been identified in suspensions of styrene-grown cells of a *Pseudomonas* sp. incubated with styrene, indicating it could be an intermediate of styrene metabolism and that these cells contained styrene monooxygenase activity [5]. Biotransformation of styrene to styrene oxide has also been reported using the methanotroph *Methylosinus trichosporon* OB3b [6] and the alkane-utilizing *Nocardia corallina* B-276 [7]. To further assess the potential of producing substituted chiral styrene oxides by biotransformation we set out to isolate styrene-degrading microorganisms bacteria.

RESULTS

The styrene-utilizing *Xanthobacter* sp. strain 124X [8] was studied first. Styrene transformation by whole cells of this strain was oxygen-dependent but we were not successful in resolving the nature of the oxidation product. Strain 124X did however contain a novel enzymic activity which isomerized styrene oxide to phenylacetaldehyde, and which was designated styrene oxide isomerase [9]. We subsequently set out to isolate other styrene-degrading organisms. As we anticipated that styrene would be toxic at higher concentrations three different isolation methods employing relatively low concentrations of styrene were used [10]. All three methods resulted in the isolation of microorganisms capable of growth with styrene as sole source of carbon and energy. 2 Fungi and 14 bacteria which appeared to be morphologically different were isolated [10].

Styrene metabolism in the new bacterial isolates was investigated. In most strains a novel FAD-dependent styrene monooxygenase activity forming styrene oxide from styrene was detected [10]. Styrene oxide accumulation was not observed due to the presence of high levels of styrene oxide isomerase activity. Based on the observed enzyme activities and the accumulation of phenylacetic acid from styrene the following degradative pathway for styrene was proposed for the styrene-degrading strain S5 [10].

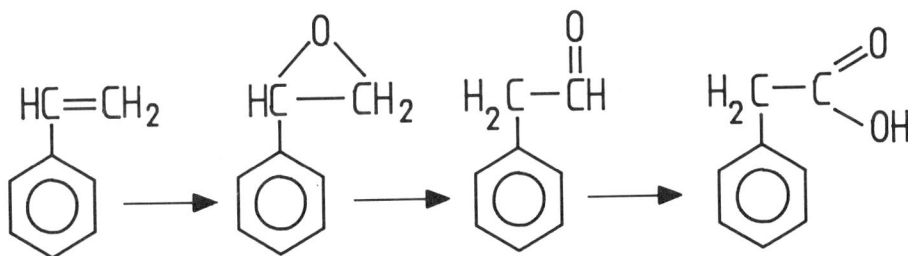

Figure 1.   Degradation pathway of styrene in strain S5.

Incubation of styrene-grown cells with α- and β-methyl styrene resulted in the accumulation of the corresponding epoxides. Determination of the enantiomeric composition of the these epoxides is in progress. Attempts to isolate mutants defective in styrene oxide isomerase are also under way. It is expected that such mutants will permit the production of a wide range of substituted styrene epoxides from the corresponding substituted styrenes. Aspects associated with the production of these epoxides which will be investigated in the near future are cofactor regeneration, product toxicity and the application of organic solvents in the production process.

REFERENCES

1. Hartmans, S., de Bont, J.A.M. and Harder, W., Microbial metabolism of short-chain unsaturated hydrocarbons. FEMS Microbiol. Rev., 1989, 63, 235-64.

2. Habets-Crützen, A.Q.H., Carlier, S.J.N., de Bont, J.A.M., Wistuba, D., Schurig, V., Hartmans, S. and Tramper, J., Stereospecific formation of 1,2-epoxypropane, 1,2-epoxybutane and 1-chloro-2,3-epoxypropane by alkene-utilizing bacteria. Enzyme Microbial Technol., 1985, 7, 17-21.

3. Weijers, C.A.G.M., van Ginkel, C.G. and de Bont, J.A.M., Enantiomeric composition of lower epoxyalkanes produced by methane-, alkane-, and alkene-utilizing bacteria. Enzyme Microbial Technol., 1988, 10, 214-18.

4. Archelas, A., Hartmans, S., and Tramper, J., Stereoselectieve epoxidation of 4-bromo-1-butene and 3-butene-1-ol with three alkene-utilizing bacteria. Biocatalysis, 1988, 1, 283-292.

5. Shirai, K. and Hisatsuka, K., Production of β-phenetyl alcohol from styrene by Pseudomonas 305-STR-1-4. Agric. Biol. Chem., 1979, 43, 1399-1406.

6. Furuhashi, K., Shintani, M. and Takagi, M., Effects of solvents on the production of epoxides by Nocardia corallina B-276. Appl. Microbiol. Biotechnol., 1986, 23, 218-23.

7. Higgins, I.J., Hammond, R.C., Sariaslani, F.S., Best, D., Davies, M.M., Tryhorn, S.E. and Taylor, F., Biotransformation of hydrocarbons and related compounds by whole organism suspensions of methane-grown Methylosinus trichosporium OB 3b. Biochem. Biophys. Res. Comm. 1979, 89, 671-77.

8. van den Tweel, W.J.J., Janssens, R.J.J., and de Bont, J.A.M., Degradation of 4-hydroxyphenylacetate by Xanthobacter 124X. Ant. van Leeuwenhoek, 1986, 52, 309-18.

9. Hartmans, S., Smits, J.P., van der Werf, M.J., Volkering, F. and de Bont, J.A.M., Metabolism of styrene oxide and 2-phenylethanol in the styrene-degrading Xanthobacter 124X. Appl. Environ. Microbiol., 1989, 55, 2850-55.

10. Hartmans, S., van der Werf, M.J., and de Bont, J.A.M., Bacterial metabolism of styrene involving a novel FAD-dependent styrene monooxygenase. In press.

ENZYME REACTIONS IN PREDOMINANTLY ORGANIC MEDIA: MEASUREMENT AND CHANGES OF pH

RAO H. VALIVETY, LINDA BROWN, PETER J. HALLING,
GRANT A. JOHNSTON AND COLIN J. SUCKLING
Departments of Bioscience & Biotechnology and Pure & Applied
Chemistry, University of Strathclyde, Glasgow G1 1XW, U.K.

ABSTRACT

Very hydrophobic esters of fluorescein with 3,7,11-trimethyl-dodecan-1-ol ionize while dissolved in pentan-3-one, in response to the pH of an equilibrated aqueous phase, though the phenolate anion remains essentially completely in the organic phase. The pH range over which they respond may be altered by bromination (4,5- or 2,4,5,7-), and is also affected by the cation concentration in the aqueous phase. These indicators may be used to measure the pH of an aqueous phase of very small volume, such as occurs in the catalyst particles when enzymes are used in mainly organic media, and which is inaccessible to other measurement methods. They show that a decrease in aqeuous pH from the pre-set value is the main reason for the lower reaction rates with short chain carboxylic acids, in esterification catalyzed by pancreatic lipase supported on celite.

INTRODUCTION

Enzymes can catalyse reactions in predominantly organic systems, which offer considerable advantages over aqueous solutions for many bioconversions [1]. The pH of the catalyst is usually adjusted to the optimal value before use [2]. Nevertheless, the effective pH of the catalyst may change during subsequent use, particularly as a result of partitioning of the reactants and/or products. Such changes have been invoked to account for some unsuccessful reactions [3-5]. However, the same sort of observations have also been explained in terms of biocatalyst specificity. The lack of an unambiguous method to determine pH of the inaccessible aqueous phase around the catalyst, which is mainly restricted to the pores of the catalyst particles, prevents proper understanding and rational design of conditions to overcome the difficulties. We now report a method to determine pH in such systems, and confirm that pH changes (rather than the intrinsic specificity of the enzyme) account for much of the reduction in rate of lipase-catalysed esterifications with short chain fatty acids.

Earlier studies of low water systems have not been able to measure unambiguously the pH in the inaccessible, relatively water-rich phase around the catalyst. Cambou & Klibanov [3] observed colour changes of a conventional pH indicator added to an aqueous-organic two-phase reaction system. However, the interpretation of these in terms of a precise pH is uncertain, because: i) the indicator would be distributed between the phases, with differing degrees of protonation in each; and ii) that in the bulk organic phase would be more visible, and its ionisation would not respond to pH in the simple way expected from the aqueous pK value. The uncertainies involved in measurements by washing the recovered catalyst in excess water have been acknowledged [5].

A more reliable method would be to use very hydrophobic indicators that remain almost completely in the organic phase in both the ionised and neutral forms, but nevertheless respond to the pH of an adjacent aqueous phase. Fluorescein esters (9-(2-carboxyphenyl)-6-hydroxy-3H-xanthine-3-one 3,7,11-trimethyl dodecyl esters) partition predominantly into the organic phase thereby enabling them to function as pH indicators in the organic phase. Their acid strength, and hence the pH range over which they respond, may be altered by bromination, giving the 4,5-dibromo or 2,4,5,7-tetrabromo derivatives [6].

Dibromo indicator dissolved in pentan-3-one is affected by the pH of an equilibrated bulk aqueous phase and the anion fraction as a function of pH is presented in Figure 1. It is pertinent to mention here that the extent of ionization does not change even over a wide range of phase volume ratios (org/aq : 1 - 400) and the indicator response is the same even when using an aqueous phase trapped entirely in the pores of celite particles (provided pH was correctly adjusted before drying and partial re-hydration). The behaviour is however not as simple as in a normal one-phase aqueous system. The curve is less steep, and it is shifted to higher pH if the aqueous Na^+ concentration is reduced. Decanoic acid, at the concentration used in the enzyme reaction, slightly shifts the curve: the other reactants have no effect. A theoretical analysis in terms of counter-ion movements, ion-pairing in the organic phase and related effects can account qualitatively for all the behaviour, and quantitatively for much of it (manuscript in preparation).

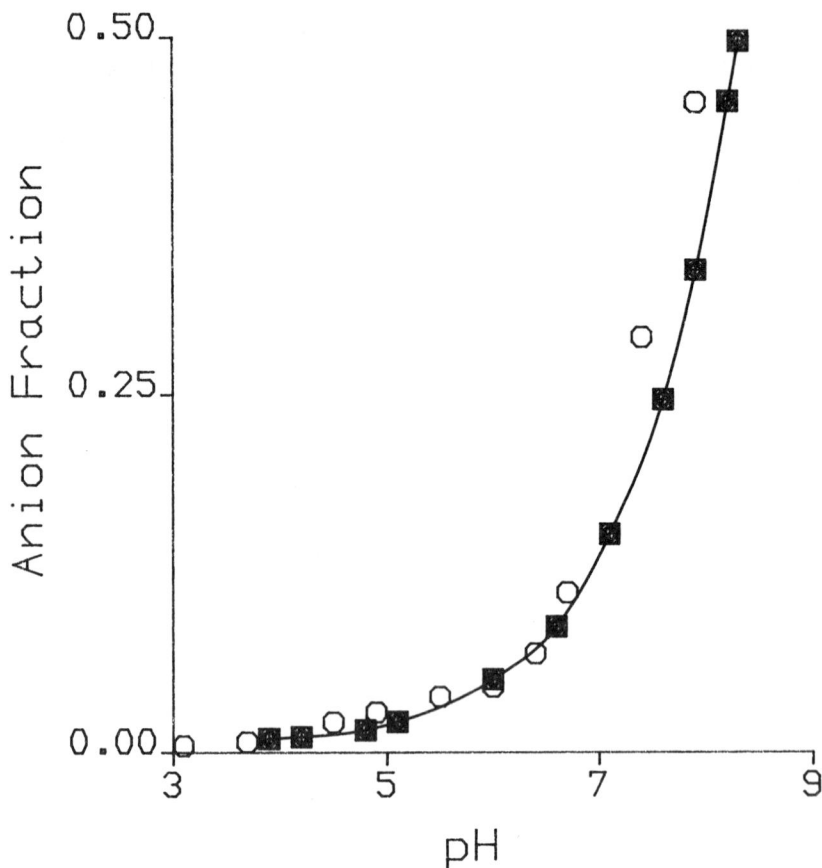

Figure 1. Effect of aqueous phase pH on ionisation of indicator equilibrated organic phase.

Aqueous phase: 0.15 M Na$_4$P$_2$O$_7$, pH adjusted with H$_3$PO$_4$. Organic phase: 2.86 X 10^{-5} M dibromo indicator in pentan-3-one (O), or containing 0.25 M decanoic acid as well (■). Equal volumes emulsified to equilibration, aqueous pH confirmed by direct measurement with glass electrode, and absorbances of organic phase samples measured. A$_{650}$ used for turbidity correction (small or zero), A$_{465}$ and A$_{530}$ used to calculate indicator acid and anion concentrations, with extinction coefficients at each wavelength.

The model system used was a celite-lipase catalyst for esterification reactions in pentan-3-one. An aqueous phase was present, entirely confined to the pores of the support, which had the appearance of a slightly moist powder. As mentioned earlier, the indicator response is affected by the cation concentration (strictly activity) in the aqueous phase, as well as pH. The aqueous $[Na^+]$ is calculated as 0.6 M (during the reaction), and hence the calibration curve at 0.6M Na^+ is employed.

TABLE 1.

Reaction rates and pH in lipase-catalysed esterifications

Acid	Esterification rate	Dibromo indicator anion fraction	Estimated pH
	μmole/s/kg catalyst		
Decanoic	9.0	0.19	7.3
Butanoic	3.32	0.147	6.8
Propanoic	0.85	0.095	6.6
Acetic	0.25	0.056	6.1

Catalyst: Hyflo Supercel celite was acid washed, suspended in 0.01 M $Na_4P_2O_7$ and pH adjusted to 7.5 with H_3PO_4. Supernatant was discarded, pancreatic lipase (Sigma L3126) stirred into slurry (0.12 g per g celite), then freeze dried. Rehydrated before use with 10% w/w water. Reaction: Catalyst (50 g/l) stirred in water-saturated pentan-3-one solution of dodecanol and carboxylic acid (both 0.25 M) at 20 °C. Organic phase samples analysed by GC. When present, indicator was 2.86 $\times 10^{-5}$M in organic phase, ionisation state determined as Fig. 1. In separate reactions with tetrabromo indicator present, this remained almost completely in anion form throughout, showing pH greater than 4.0. Esterification rates were not affected by the presence of either indicator.

The initial rates of esterification with acids of various chain lengths are presented in Table 1. The reaction rate declined sharply with the short chain acids. The pH of the aqueous phase around the catalyst, determined employing our novel technique, was found to be significantly reduced with the short chain acids. A separate celite-lipase catalyst was prepared with the pH adjusted in advance to 6.1, the value found in the presence of acetic acid. In a reaction with decanoic acid, under the same conditions as before, this catalyst gave a rate of 1.08 μmole/s/kg catalyst. This finding confirms that pH was responsible for much of the reduction in rate with acetic acid.

There remain some limitations to the pH measurement technique described here. Firstly, it fails in the presence of basic reactants that can form organic-soluble cations. Secondly, the current indicators cannot be used with less polar solvents. Thirdly, we cannot confidently predict the behaviour in reaction systems where thermodynamic water activity is substantially less than 1: these also pose a theoretical problem in defining pH. Current work is aimed at overcoming these limitations.

Our method for pH measurement in these systems allows us to determine to what extent pH shifts are responsible for observed poor reaction rates. This knowledge provides a rational basis for the selection of better conditions.

ACKNOWLEDGEMENTS

We thank SERC Biotechnology Directorate and ICI, Biological Products for financial support.

REFERENCES

1. Latest reviews: ; Klibanov, A.M., <u>Trends Biochem. Sci.</u>, 1989, **14**, 141 ; Dordick, J.S., <u>Enzyme Microb. Technol.</u>, 1989, **11**, 194

2. Discussed in detail by: Zaks, A. and Klibanov, A.M., <u>J. Biol. Chem.</u>, 1988, **263**, 3194.

3. Cambou,B. and Klibanov, A.M., <u>Biotechnol. Bioeng.</u>, 1984, **26**, 1449.

4. Abraham, G., Murray, M.A. and John, V.T., Biotechnol. Lett., 1988, **10**, 555.

5. Cassells, J.M. and Halling, P.J., <u>Biotechnol. Bioeng.</u>, 1989, **33**, 1489.

6. Syntheses: Suckling, C.J., Halling, P.J., Johnston G.A. and Brown, L., UK Pat. Appl., 1990, 88 24145.

THE BIOTECHNOLOGICAL POTENTIAL OF <u>ESCHERICHIA COLI</u> TYPE B STRAIN SPAO TRANSAMINASES FOR L-METHIONINE SYNTHESIS.

N.F. Shipston and A.W. Bunch.

The Biological Laboratory, The University, Canterbury, Kent CT2 7NJ, U.K.

BACKGROUND

Methionine is one of the most important of the commercially manufactured amino acids. The chemical route for methionine synthesis is a multistage process with the main carbon skeleton originating from acrolein. Resolution of the two optical isomers generated is performed in the latter stages and usually involves biological technology requiring several steps in the procedure. There are several biological systems which have the potential to improve the efficiency and yield of the L-isomer of methionine, although the chemical synthetic route usually requires modification. However L-amino acid dehydrogenases, for example, require cofactors which have to be recycled by other enzyme processes (or chemical methods). In contrast transaminases do not need cofactor recycle facilities. Additionally, microbial L-amino acid transaminases are comparatively stable and could be used as cell-free catalysts on their own. Although they require pyridoxal phosphate for activity this cofactor is usually tightly bound and therefore not easily lost. One final characteristic of these enzymes which makes them attractive for biotransformations is their apparently broad substrate specificity.

Microbial transaminases have received limited attention in comparison to many other enzymes. The reactions they catalyse are usually freely reversible and stereoselective. Transaminases that produce D-amino acids are known but less common. Unlike many amino acids L-methionine does not require a specific transaminase reaction for its synthesis as it is a product derived from aspartic acid[1]. Nevertheless, methionine transaminases exist, but their significance in methionine metabolism is still being assessed[2]. For many years it has been known that E. coli strains have the ability to synthesise ethylene. The first step in the biosynthetic pathway involved has been shown to involve a methionine transaminase.

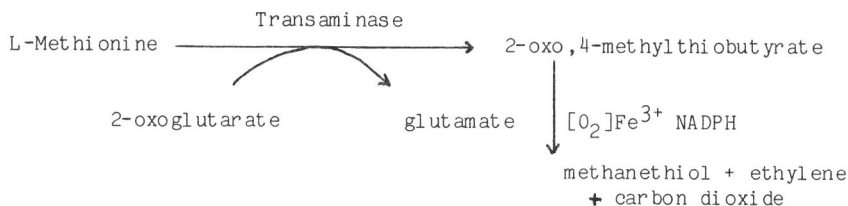

L-Methionine $\xrightarrow{\text{Transaminase}}$ 2-oxo,4-methylthiobutyrate

2-oxoglutarate   glutamate $\Big\downarrow [O_2]Fe^{3+}$ NADPH

methanethiol + ethylene + carbon dioxide

Ethylenogenic capabilities from methionine can be used as a simple screen for methionine transaminase activity in microorganisms[3]. The most ethylenogenic microbial strain so far isolated is E. coli typ B SPAO, the methionine transaminase potential of which is the subject of this paper. E. coli strains can utilise up to nine amino acids as sole sources of carbon and nitrogen[4]. Cell-free extracts of these bacteria have been shown to catalyse the in vitro transamination of nineteen L-amino acids with glutamate as the amino donor[5].

Strains of E. coli have been shown to possess at least nine distinct amino transferases[6]. Many have been purified to homogeneity and their primary structures sequenced and it has been shown that several of these transaminases have overlapping substrate specificities. This has hindered studies on these enzymes using powerful techniques such as gene mutation. It has been shown that the products of the tyr B and asp C genes have L-methionine transaminase activity as well as L-phenylalanine/L-tyrosine and L-aspartic acid activities respectively[7]. The properties of these two enzymes are very similar. The regulation of L-methionine transaminase activity in E. coli has as yet been poorly characterised.

Our objective was to investigate the nature of the methionine transaminase activity in ethylenogenic strains of E. coli typ B SPAO in order to determine whether this bacterium possessed a novel transaminase for use in L-methionine synthesis.

OBSERVATIONS

Ince and Knowles put forward the hypothesis that ethylenogenesis was a means of acquiring the amino-nitrogen from L-methionine in E. coli SPAO[8]. Subsequent work has shown that the ammonium concentration and carbon to nitrogen ratio in culture media effects the amount of ethylene synthesised from methionine via a transaminase reaction[9].

We have shown that over 85% of the methionine catabolic activity that results in the production of 2-oxo, 4-methylthiobutric acid in E. coli SPAO is associated with a transaminase. This activity can be abolished by the addition of hydroxylamine which chemically reacts with pyridoxal phosphate. No L-methionine-γ-lyase activity has been detected in cell-free extracts of this bacterium although small amounts of 2-oxobutyrate have been detected in culture supernatent. Controls have shown that over the time scale of our assays very little (<2%) of the methionine degrades chemically to 2-oxo, 4-methylthiobutyric acid.

Transaminase activity is proportional to protein concentration, and has a pH optimum close to 7.8 at 37°C. Several 2-oxoacids could replace 2-oxoglutarate as an amino accepter as shown in Table 1. However, the presence of methionine in culture media appears to lower the specific activity of the methionine transaminase(s) (Table 2).

Table 1
Co-substrate specificities of L-methionine transaminase in E. coli type B SPAO.

Substrate	Relative specific activity
2-oxoglutarate	100
2-oxo,4-methylthiobutyrate	107
2-oxobutyrate	106
oxaloacetate	99
pyruvate	91
glyoxylate	67

Table 2
Specific activities of L-methionine transamination in High Speed
Supernatant derived from cells grown in defined and complex media.

Growth Medium	Specific Activity
	(μmol h^{-1} mg protein^{-1})
Defined	1.13
Defined + L-methionine	0.90
Nutrient broth	0.39

Several amino acids have the capability of lowering the conversion of radioactive methionine to labelled 2-oxo, 4-methylthiobutyrate. The most potent of these are L-aspartate (25% inhibition), L-tyrosine (34% inhibition) and L-phenylalanine (52% inhibition). It is noteworthy given what is known about the transaminases of E. coli that L-isoleucine, L-leucine, and L-valine caused little inhibition. D-methionine did not act an inhibitor of the process. Analysis of the inhibition kinetics for L-tyrosine revealed that it acted as a competitive inhibitor. No clear indication of such a characteristic was revealed from analysing the results obtained with aspartate. Both tyrosine and aspartate transaminase activity are present in cell-free extracts of E. coli type B SPAO. The aspartate transaminase was stable at 55°C, when held at this temperature for over 60 minutes. Both tyrosine and methionine transaminases were inactivated rapidly by incubation at 55°C. Finally the K_m of the transaminase for methionine is approximately 15mM, in the presence of a saturating concentration of 2-oxoglutarate (5mM).

We have used several purification protocol in an attempt to resolve the methionine, tyrosine and aspartate acid transaminase activities. In all the procedures so far used we have been able to separate the aspartate transaminase activity from the other two activities, but never L-tyrosine transaminase from the L-methionine transaminase (for example see Figure 1).

Physiological studies have shown that the presence of L-tyrosine in growth media can greatly reduce the methionine to 2-oxo, 4-methylthiobutyric transformation and overall ethylene synthesis.

Figure 1 Hydroxyapatite Column Chromatography of L-Methionine, L-Tyrosine and L-Aspartate Aminotransferase Activity of E. Coli B SPAO

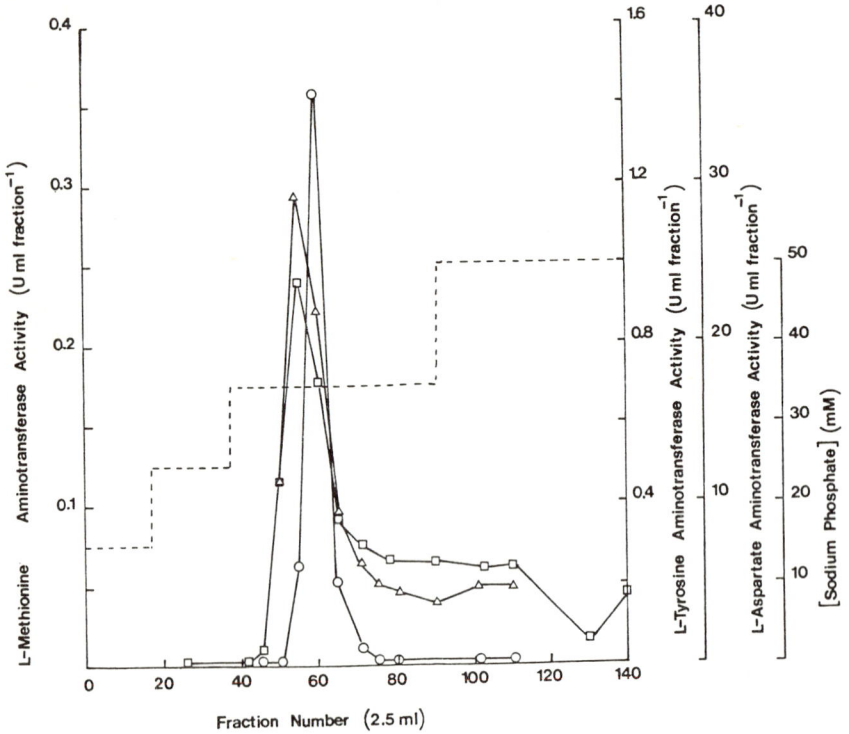

Concentrated, dialysed protein (90mg) containing pooled, active fractions from anion exchange chromatography was loaded onto the column (Biogel HT; 1.6 x 15cm) and eluted step-wise with 7.5, 12.5, 17.5 and 25mM sodium phosphate (pH 7.2) buffer at 30ml h^{-1}. Fractions (2.5ml) were collected and aminotransferase activity towards L-methionine ( ), L-tyrosine (Δ and L-aspartate ( ) was measured.

Cultures of E. coli K12 are able to convert much less methionine to ethylene than E. coli type B SPAO. However when we examined a mutant of this bacterium which had both the tyrosine and aspartic acid transaminase structural genes deleted, ethylene synthesis from methionine could no longer be detected. All these results indicate that the methionine transaminase and tyrosine transaminase activities are associated with the same gene product, although this is being investigated further using both genetic and biochemical techniques.

Our results therefore imply that it is unlikely that E. coli type B SPAO has a novel methionine transaminase activity. Therefore we will develop the biotechnological uses of methionine transaminases by using E. coli K12 where sequence data and more developed molecular biology techniques are available. (All details of media and methodology used in the investigations outlined above can be found in reference 10.)

Future Work

At this time we are in the process of isolating the tyrosine transaminase gene from E. coli K12 as a first step in the development of high expression strains. Both E. coli K12, type B SPAO and strains of Lactococcus cremoris are being assessed as candidates for operation of the transaminase activity. The latter bacterium has been chosen with a view the possible exploitation of the genetically engineered strain in membrane bioreactors. Further strain development will involve alteration of the basic gene structure to enhance catalytic efficiency and substrate specificity. This will allow higher conversion yields and the possible generation of novel amino acid products.

Acknowledgements

We would like to thank the following: Professor C.J. Knowles for many helpful discussions. The SERC for financial support. Miss Siobhan McSwiggan for the data on the mutant E. coli K12 strains and Dr. Paul Taylor for the mutant of E. coli. Finally, Ms Sue Daives and Ms Mary-Clare Mirwald for typing the manuscript.

References

1. Umbarger, J.E. (1978) Amino acid biosynthesis and its regulation. Annual Reviews of Biochemistry 47, 533-606.

2. Collier, R.H. and Dohlaw, G. (1972) Non-identity of the aspartate and the aromatic aminotransferase components of transaminase A in Escherichia coli. Journal of Bacteriology 112, 365-371.

3. Mansouri, S. and Bunch, A.W. (1989) Bacterial ethylene synthesis from 2-oxo, 4-thiobutyric acid and from methionine. Journal of General Microbiology 135, 2819-2827.

4. Halvorson, G. (1972) Utilization of single L-amino acids as sole source of carbon and nitrogen by bacteria. Canadian Journal of Microbiology 18, 1647-1650.

5. Chesne, S. and Pelmont, J. (1973) Glutamate oxaloacetate transaminase d'Escherichia coli I. Purificatin et Specificite. Biochemie 55, 237-241.

6. Mavrides, C. (1987) Transamination of aromatic amino acids in Escherichia coli. In: Methods in Enzymology, 142 (ed. S. Kaufman) pp. 253-257. Academic Press, London.

7. Powell, J.T. and Morrison, J.F. (1978) The purification and properties of the aspartate aminotransferase and aromatic amino transferase from Escherichia coli. European Journal of Biochemistry 87, 391-400.

8. Ince, J.E. and Knowles, C.J. (1986) Ethylene formation by cell extracts of Escherichia coli. Archives of Microbiology 146, 151-158.

9. Shipston, N.F. and Bunch, A.W. (1989) The physiology of L-methionine catabolism to the secondary metabolite ethylene by Escherichia coli. Journal of General Microbiology 135, 1489-1497.

10. Shipston, N.F. (1990) Physiology and enzymology of L-methionine catabolism to the secondary metabolite ethylene in Escherichia coli. Ph.D. thesis, The University of Kent at Canterbury.

STREPTOMYCES GRISEOLUS REGIOSPECIFIC O-DEALKYLATION OF 4-METHOXYPHENYL SULFONYLUREAS

BARRY STIEGLITZ AND MARTIN T. SCOTT
Agricultural Products Department, Experimental Station
E. I. du Pont de Nemours & Company
Wilmington, Delaware 19880-0402, U.S.A.

ABSTRACT

Streptomyces griseolus ATCC 11796 regiospecifically O-dealkylated two 4-methoxyphenyl sulfonylurea herbicides to their corresponding 4-hydroxyphenyl sulfonylureas. No other O-dealkylated sulfonylureas were formed. These bioconversions were scaled-up in small fermentors and the 4-hydroxyphenyl sulfonylureas were purified by preparative high performance liquid chromatography. The purified 4-hydroxyphenyl sulfonylureas can be used as analytical standards in sulfonylurea plant and animal metabolism studies.

INTRODUCTION

Sulfonylureas (SU's) are a class of herbicides with the general structure shown in Figure 1.

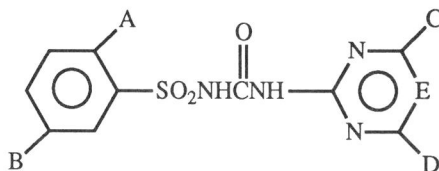

Figure 1. General Structure of a SU Herbicide

A, B, C, and D represent various substituents and E may be N or CH.

Previous studies by Romesser and O'Keefe showed that *Streptomyces griseolus* ATCC 11796 performed a variety of co-metabolic transformations on several SU herbicides including O-dealkylation of chlorosulfuron and chlorimuron ethyl, the active ingredients of Glean® and Classic® herbicides, respectively (1). Both SU's contain a methoxy substituent in the C/D positions.

The objective of this study was to generate via regiospecific O-dealkylations shown in Figure 2, the two 4-hydroxyphenyl sulfonylureas, G7460 and 33133 from their corresponding 4-methoxyphenyl sulfonylureas, B8346 and T8047.

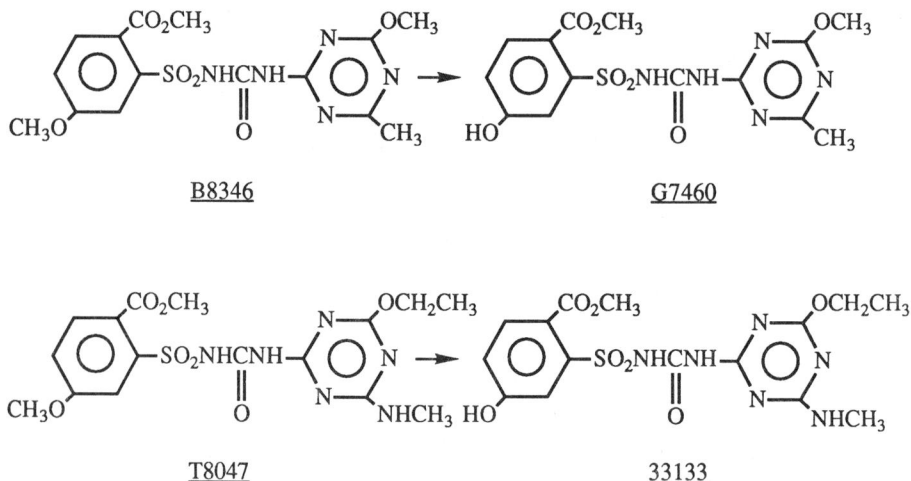

Figure 2. Regiospecific O-dealkylation of B8346 and T8047

Both G7460 and 33133 are plant metabolites of metsulfuron methyl (2) and DPX-A7881 (M. T. Scott, private communication) the active ingredients of Ally® and Muster® herbicides, respectively. In this study, we describe the *Streptomyces griseolus* ATCC 11796 regiospecific O-dealkylation of B8346 and T8047 and the preparation of mg quantities of G7460 and 33133 by coupling small fermentor bioconversions with preparative high performance liquid chromatography (HPLC).

MATERIALS AND METHODS

The chemical purity of B8346 and T8047 was 99% and 95% respectively as determined by HPLC. Solvents for chromatography were HPLC grade and all other chemicals and materials were of the highest purity available.

Microbial Transformation

Microbial transformations of B8346 and T8047 were carried out with the bacterium *Streptomyces griseolus* ATCC 11796 which was grown and induced in a two-stage process for SU biotransformations as described previously (1). Microbial transformations were performed with a gyratory shaker in 125 ml Erlenmeyer flasks containing 25 ml of sporulation medium (1) and 120 ppm SU or in a 3 liter Bioengineering AG fermentor

(Model KLF 2000) containing 2 liters of sporulation medium and 300-350 ppm SU. The latter was added in two equal portions at 4 hr and 20 hr following inoculation. Cultures were harvested at 24 hr (shake flasks) and 68 hr (fermentor) by centrifugation and filtration, respectively. Cell-free broths were acidified to pH 2.0 and extracted with 2 volumes of diethyl ether which was evaporated to dryness with a stream of nitrogen (shake flask extract) or in a rotary evaporator (fermentor extract). Residues were dissolved in acetonitrile:water for HPLC analysis and purification.

Analytical Methods

HPLC was accomplished with a Perkin Elmer Series 4 Liquid Chromatograph system including a UV/visible spectrophotometer detector (Model LC-95 set at 230 nm). Metabolites in shake flask extracts were separated and purified by collecting individual fractions on a Zorbax ODS (Du Pont Co.) semi-preparative column (9.4 mm x 25 cm) using a 30 min linear gradient of acetonitrile-water-phosphoric acid from 10:90:0.1 to 75:25:0.1 at a flow rate of 2.5 ml/min. the acetonitrile in metabolite fractions was removed by rotary evaporation and followed by solvent extraction as described above. The isolated metabolites were analyzed by mass spectrometry (MS).

Preparative HPLC was carried out with a Rainin Rabbit™ system (HPX pumps, Apple Macintosh Plus™ computer with Rainin MacRabbit™ software; Rainin Instrument Co.), a Knauer UV detector (Model 87) set at 254 nm and a Rainin™ Dynamax column (C_{18}, 41.4 mm x 25 cm). B8346 metabolites were separated with a 50 min gradient of acetonitrile-water-phosphoric acid from 45:55:0.1 to 55:45:0.1 at a flow rate of 45 ml/min. T8047 metabolites were separated with a 65 min linear gradient of acetonitrile-water-phosphoric acid from 35:65:0.1 to 55:45:0.1 at a flow rate of 45 ml/min. Metabolites were recovered from fractions as described above.

All mass spectrometry analyses were carried out by positive ion thermospray HPLC/MS on a Finnigan Model 4600 mass spectrometer. Proton nuclear magnetic resonance (NMR) were recorded using a Varian Model XL-200 MHZ Instrument.

RESULTS AND DISCUSSION

HPLC analysis of shake flask biotransformations with either B8346 or T8047 showed one major and minor metabolite which were more polar than the parent SU. The major metabolites derived from either B8346 or T8047 were estimated to be formed in 80% yield. Purification of these major metabolites and subsequent thermospray HPLC/MS analysis gave molecular and fragment ions corresponding to G7460 (B8346 derived) and

33133 (T8047 derived). Control flasks (no cells) incubated with either B8346 or T8047 in sterile biotransformation medium gave no metabolites.

To generate larger quantities of G7460 and 31333, 2 liter fermentor bioconversions were carried out with 300 ppm B8346 and 350 ppm T8047. Approximately 900 mg and 650 mg of crude biotransformation extract was recovered from the B8346 and T8047 fermentor broths, respectively. Figure 3A represents the chromatogram of 300 mg of crude B8346-derived material (0.5 mg/ml in 30% acetonitrile) loaded onto the preparative HPLC column. The purified peaks at 21 min and 27.5 min were compared by co-chromatography and shown to be G7460 and B8346, respectively. G7460 identification was confirmed by NMR analysis. The fractionated G7460 was 95% pure by HPLC area percent and the recovered yield (mg G7460 ÷ mg B8346 added x 100) was 52%. No other hydroxylated product was detected.

Figure 3B shows a chromatogram representing 250 mg of crude T8047-derived extract (0.5 mg/ml in 30% acetonitrile) loaded onto the preparative HPLC column. The large peak at 27 min included 33133 and two impurities. Following further preparative HPLC fractionation, 95% pure 33133 was obtained and proton NMR confirmed its 4-hydroxyphenyl structure. The recovered yield of 33133 was 31%. HPLC/MS of 33133 impurities and other minor peaks showed no other O-dealkylated SU.

These results indicate that _Streptomyces griseolus_ ATCC 11796 can regiospecifically O-dealkylate B8346 and T8047 to G7460 and 33133, respectively. Also these microbial transformation can be scaled-up in small fermentors and metabolites purified by preparative HPLC. Therefore, we can prepare sufficient quantities of 4-hydroxyphenyl SU's for use as analytical standards in plant and animal metabolism studies required for registration of new agrichemicals. Furthermore, both G7460 and 33133 can be used as intermediates for the synthesis of their glucose conjugates which are also plant metabolites (2) (M. T. Scott, private communication). These regiospecific O-dealkylations represent another example of using microbes to generate metabolites that are often difficult to prepare by chemical synthesis.

99

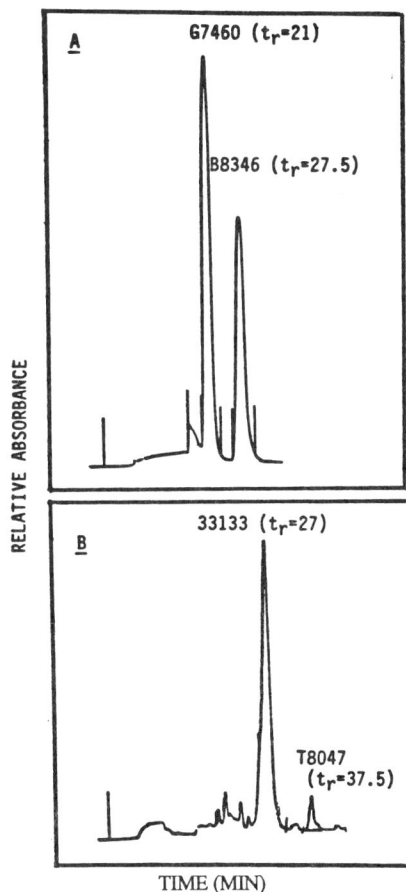

Figure 3. Preparative HPLC Chromatograms of B8346 (A) and
T8047 (B) Crude Bioconversion Extracts

ACKNOWLEDGMENTS

We thank L. M. Shalaby for obtaining MS spectra and D. D. Stranz for obtaining the
NMR spectra of metabolites. We also thank D. J. Marquis-Omer, D. A. Suchanec and
L. R. Proksch for technical assistance and A. M. Strickland for preparing the manuscript.

REFERENCES

1. Romesser, J. A. and O'Keefe, D. P., Induction of cytochrome P-450-dependent
   sulfonylurea metabolism in _Streptomyces griseolus_. Biochem. Biophys. Res. Comm.,
   1986, 140, 650-659.

2. Beyer, E. M., Duffy, M. J., Hay, J. V., and Schlueter, D. D., Sulfonylurea Herbicides.
   In Herbicides: Chemistry, Degradation and Mode of Action, Vol. 3, ed., P. C. Kearney
   and D. D. Kaufman, Marcel Dekker Inc., New York 1988, pp. 117-89.

ENANTIOSELECTIVE ENZYMATIC ACYLATION USING ACID ANHYDRIDES

B.BERGER, K.KÖNIGSBERGER, K.FABER and H.GRIENGL

Institute of Organic Chemistry and Christian Doppler Laboratory
for Chiral Compounds, Graz University of Technology,
Stremayrgasse 16,A-8010 Graz, Austria

ABSTRACT

Enzymatic resolution of racemic bicyclic alcohol *rac*-1a was accomplished using lipase AY-30 from *Candida cylindracea* as biocatalyst and acetic anhydride as acyl donor. The enantioselectivity of the enzyme was found to depend strongly on the reaction conditions employed: Whereas low selectivity (E = 18) was observed without precautions taken in order to remove the produced acetic acid, an efficient resolution was obtained by addition of inorganic or organic base (E > 200). Alternatively, adsorption of the biocatalyst onto celite served even better for this purpose, enhancing E to about 370.

INTRODUCTION

Enzyme catalysed acylation in organic media has been shown to be advantageous over hydrolytic reactions in particular due to the following reasons:

i) Possible improvement of enantioselectivity of the enzyme [1][2][3],

ii) Successful transformation of substrates being poorly soluble in aqueous systems [4],

iii) better overall yields since loss-causing extractive workup is omitted,

iv) lack of undesired side-reactions requiring water [5],

v) enzymes can be recovered by simple filtration since they are totally insoluble in lipophilic organic media and

vi) there is no risk of microbial contamination.

In order to avoid the unfavourable equilibrium situation in trans- and interesterification reactions [1], special acyl donors making the acyl-transfer reaction completely irreversible have recently been employed:

    a)   Enol esters [6], such as vinyl acetate,

    b)   oxime esters [7] and

    c)   acid anhydrides [3].

    Whereas the first of these methods has already gained widespread use [8], the limited availablilty of oxime esters represents a serious drawback. Acid anhydrides, however, can readily be used here as cheap acyl donors for enzyme-catalysed esterifications. All of these methods have in common that they produce one equivalent of unavoidable by-product:

    a) an aldehyde/ketone, b) an oxime or c) a carboxylic acid.

Aiming to elucidate possible side-effects of the acid produced in method c) we investigated the enzyme-catalysed resolution of the tetrachloro bicyclo[2.2.1]heptane system shown below. With this particular substrate the corresponding hydrolysis failed due to the complete insolubility of the substrate acetate (*rac-1b*) in water. As shown in scheme 1, both enantiomers of this compound can be used as building blocks for the synthesis of prostaglandins [9] [10] and functionalized carbocyclic nucleoside analogues [11].

SCHEME 1

Synthesis of biologically active compounds.

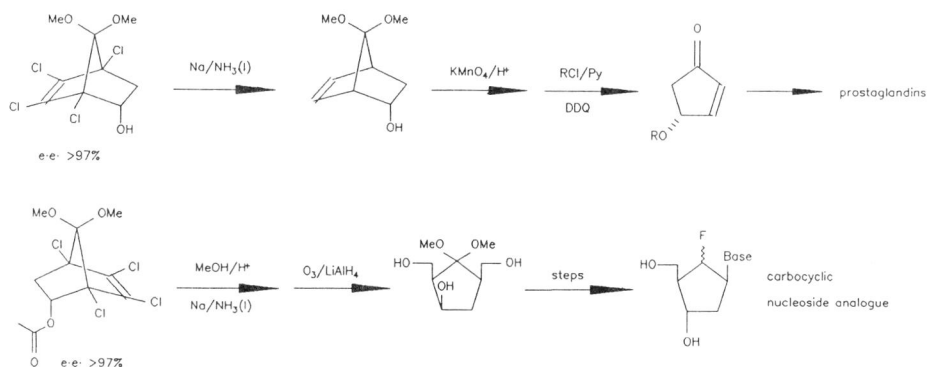

RESULTS AND DISCUSSION

As depicted in scheme 2, lipase [12] catalysed acylation of bicyclic starting alcohol *rac*-**1a** using acetic anhydride in toluene proceeded with low selectivity (E = 18) [13]. Addition of suspended inorganic or dissolved organic base, however, improved E more than ten-fold. Best results were obtained when the biocatalyst was adsorbed onto celite [3], leading to an enantiomeric ratio of 370. By this way both enantiomers (1R,2S,4S)-**1a** and (1S,2R,4R)-**1b** can be obtained in >97 % e.e. when the reaction is terminated at 50 % conversion.

SCHEME 2

Enzymatic Resolution of *rac*-**1a**.

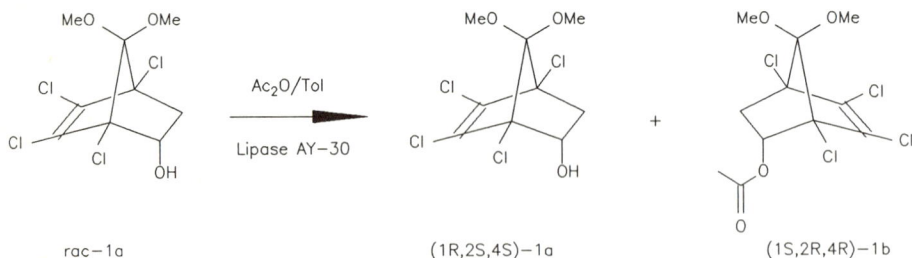

rac—1a   (1R,2S,4S)—1a   (1S,2R,4R)—1b

TABLE 1

Optical Purity of Products

Base	Lipase	Conversion [%]	e.e. of 1a [%]	e.e. of 1b [%]	E [13]
none	neat	54	87	74	18
KHCO$_3$	neat	45	80	98	210
KHCO$_3$/18-crown-6	neat	41	47	66	7
2,6-lutidine	neat	47	86	97	210
none	on celite	44	80	99	370

An attempt to increase the moderate reaction rate of the heterogeneous KHCO$_3$-system by adding 18-crown-6 led to a significant drop in selectivity (E = 7), caused by concomitant chemical acylation catalysed by solubilized bicarbonate as proven *via* an independent experiment in the absence of enzyme. In all of the other systems used, no chemical acylation - a prerequisite for a high e.e. of products - could be observed [14].

EXPERIMENTAL

The optical purity of **la** was determined by GLC-analysis of the corresponding diastereomeric carbonates after derivatisation with (-)-menthyl chloroformate [15] (J&W DB 1701 column, 30m x 0.25mm, 0.25 μm film, N_2, α = 0.975). The absolute configuration was elucidated by dehalogenation [16] ($Na/NH_{3(1)}$) to yield *endo*-7,7-dimethoxybicyclo-[2.2.1]hept-5-en-2-ol with known configuration [17].

General Procedure:
To a solution of *rac*-**la** (15 mmol) and acetic anhydride (15 mmol) in toluene (30 mL) were added base (15 mmol) and lipase AY-30 [12] (2.3 g). The suspension was shaken at 200 rpm (25°C) and the reaction (monitored by GLC) was terminated when an appropriate degree of conversion was reached. Inorganics were removed by extraction with NH_4Cl solution and after column chromatography alcohol **la** and acetate **lb** were obtained in >90 % overall yield.

(1*R*,2*S*,4*S*)-7,7-dimethoxy-1,4,5,6-tetrachlorobicyclo[2.2.1]hept-5-en-2-ol (**la**): mp. 85-7°C, $[\alpha]_D^{20}$ -34.9 (c 2.54, MeOH), e.e. 98%.

(1*S*,2*R*,4*R*)-7,7-dimethoxy-1,4,5,6-tetrachlorobicyclo[2.2.1]hept-5-en-2-yl acetate (**lb**): mp. 75-7°C, $[\alpha]_D^{20}$ +47.6 (c 2.85, MeOH), e.e. 99%.

ACKNOWLEDGEMENTS

The authors wish to express their cordial thanks to Amano Pharm. Co. (Japan) for their generous gift of lipase. Financial support from Fonds zur Förderung der wissenschaftlichen Forschung and the Christian Doppler Ges. (Austria) is gratefully acknowledged.

REFERENCES AND NOTES

1. Chen, C.-S., Wu, S.-H., Girdaukas, G. and Sih, C.J., Quantitative analyses of biochemical kinetic resolution of enantiomers. 2. Enzyme-catalyzed esterifications in water-organic solvent biphasic systems. J.Am.Chem.Soc. 1987, **109**, 2812-2817.
2. Yamamoto, K., Nishioka, T., Oda, J. and Yamamoto, Y., Asymmetric ring opening of cyclic acid anhydrides with lipase in organic solvents. Tetrahedron Lett. 1988, **29**, 1717-1720.

3. Bianchi, D., Cesti, P. and Battistel, E., Anhydrides as acylating agents in lipase-catalyzed stereoselective esterification of racemic alcohols. J.Org.Chem. 1988, 53, 5531-5534.

4. Snijder-Lambers, A.M., Doddema, H.J., Grande, H.J. and van Lelyveld, P.H., Log P as a hydrophobicity index for biocatalysis; cofactor regeneration during enzymatic steroid oxidation in organic solvents. In: Studies in Organic Chemistry, ed., C. Laane, J. Tramper and M.D. Lilly, Elsevier, Amsterdam 1987, vol. 29, pp. 87-95.

5. Kazandjian, R.Z. and Klibanov, A.M., Regioselective oxidation of phenols catalyzed by polyphenol oxidase in chloroform. J.Am.Chem.Soc. 1985, 107, 5448-5450.

6. Degueil-Castaing, M., De Jeso, B., Drouillard, S. and Maillard, B., Enzymatic reactions in organic synthesis: 2-Ester interchange of vinyl esters. Tetrahedron Lett. 1987, 28, 953-954.

7. Ghogare, A. and Kumar, G.S., Oxime esters as novel irreversible acyl transfer agents for lipase catalysis in organic media. J.Chem.Soc., Chem.Commun. 1989, 1533-1535.

8. For a representative example see: Wang, Y.-F., Lalonde, J.J., Momongan, M., Bergbreiter, D.E. and Wong, C.-H., Lipase-catalyzed irreversible transesterifications using enol esters as acylating reagents: preparative enantio- and regioselective syntheses of alcohols, glycerol derivatives, sugars and organometallics. J.Am.Chem.Soc. 1988, 110, 7200-7205.

9. Jung, M.E., New approaches to the total synthesis of biologically active natural products. In: Current Trends in Organic Synthesis, ed., H.Nozaki, Pergamon, Oxford 1983, pp. 61-70.

10. Kitamura, M., Kasahara, I., Manabe, K., Noyori, R. and Takaya, H., Kinetic resolution of racemic allylic alcohols by BINAP-Ruthenium(II)-catalyzed hydrogenation. J.Org.Chem. 1988, 53, 708-10; and refs. cited therein.

11. Baumgartner, H., Bodenteich, M., Yang, S. and Griengl, H., in preparation.

12. Lipase AY-30 from Candida cylindracea was a generous gift from Amano Pharm.Co. (Japan).

13. Chen, C.-S., Fujimoto, Y., Girdaukas, G. and Sih, C.J., Quantitative analyses of biochemical kinetic resolutions of enantiomers. J.Am.Chem.Soc. 1982, 104, 7294-7299.

14. A full paper on an extended study is in preparation.

15. Westley, J.W. and Halpern, B., The use of (-)-menthyl chloroformate in the optical analysis of asymmetric amino and hydroxyl compounds by gas chromatography. J.Org.Chem. 1968, 33, 3978-3980.

16. Paquette, L.A., Learn, K.S., Romine, J.L. and Lin, H.-S., Molecular recognition during 1,2-addition of chiral vinyl organometallics to chiral β,γ-unsaturated ketones. Case studies of three 7,7-disubstituted 2-norbornenones. J.Am.Chem.Soc. 1988, 110, 879-890.

17. Königsberger, K., Faber, K., Marschner, Ch., Penn, G., Baumgartner, P. and Griengl, H., Enzymatic resolution of endo-bicyclo[2.2.1]hept-2-yl butyrates and related compounds: Steric requirements in the bridge-region. Tetrahedron 1989, 45, 673-680.

WHOLE CELL CATALYSED RESOLUTION OF A

RACEMIC BICYCLIC LACTAM

− A NOVEL APPROACH TO THE PRODUCTION OF CHIRAL CARBOCYCLIC NUCLEOSIDES

* STEVE TAYLOR, + ALAN SUTHERLAND, * CAROL LEE, * RICHARD WISDOM, * CHRISTOPHER EVANS, + STANLEY ROBERTS AND #* STEVE THOMAS.

+ EXETER UNIVERSITY CHEMISTRY DEPT, STOCKER RD, EXETER

* ENZYMATIX LTD, CAMBRIDGE SCIENCE PARK, MILTON ROAD, CAMBRIDGE

TO WHOM CORRESPONDANCE SHOULD BE ADDRESSED

ABSTRACT

A process has been defined for the resolution of the racemic lactam (±)-2-Azabicyclo[2.2.1]hept-5-en-3-one, a versatile intermediate in the synthesis of novel carbocyclic nucleosides. Both optical forms of the lactam have been obtained in very high optical purity (>98% e.e.) in a rapid, facile scaleable biotransformation process using whole cell catalysts.

INTRODUCTION

Carbocyclic analogues of purine and pyrimidine nucleosides, in which a methylene group replaces the oxygen atom of the ribofuranose ring have generated great interest as potential anti-viral agents in recent years.

One such carbocyclic nucleoside, (±) Carbovir, (I) has proven to be a potent and selective inhibitor of HIV-1 in vitro. Indeed, its hydrolytic stability and ability to inhibit infectivity and replication of the virus in human T-cell lines at concentrations 200-400 fold below toxic levels makes carbovir an excellent candidate for development as a potential anti-retroviral agent.

Studies undertaken at the laboratories of Glaxo Group Research have resulted in the characterisation of optically pure (-) Carbovir via a synthetic route beginning with the chiral natural product Aristeromycin (II). Subsequent

Carbovir

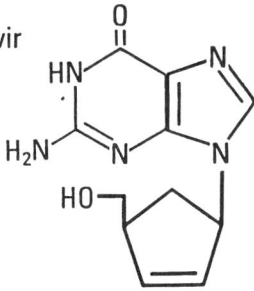

(I)

Aristeromycin

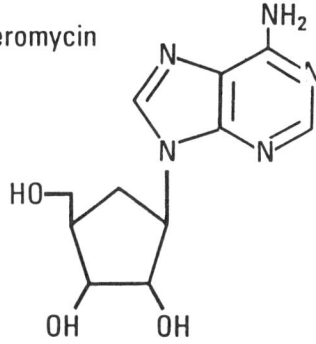

(II)

2-Azabicyclo[2.2.1]hept-5-en-3-one (γ-Lactam)

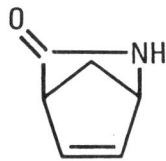

(III)

biological evaluation of (−) Carbovir in whole cell assays using MT-4 cells demonstrated that (−) Carbovir has similar activity against HIV (RF strain) to AZT (zidovudine) ($I.C._{50}$ = 0.0015 μg ml^{-1} and 0.001 μg ml^{-1} respectively).

Whilst this work demonstrated for the first time the efficacy of a single enantiomer of Carbovir _in vitro_ the synthetic route used restricts access to the (−) optical form of the drug and centres on the use of a very expensive synthon, Aristeromycin as the start point for the synthetic scheme.

More recently, an efficient synthesis to (±) Carbovir has been described by Vince and Hua (see figure 1). The key intermediate in the route is a racemic bicyclic lactam (±)-2-Azabicyclo[2.2.1]hept-5-en-3-one (III) which is available by addition of tosylcyanide to cyclopentadiene followed by acid work-up.

This work describes the use of two distinct whole cell biocatalysts in the resolution of the racemic lactam to generate both enantiomers at very high optical purities (>98% e.e.).

The technology enables both chiral forms of a wide range of potential anti-viral agents, including carbovir, to be characterised for efficacy and toxicity, in the absence of the other optical form.

Figure 1

Synthesis of
(±) Carbovir

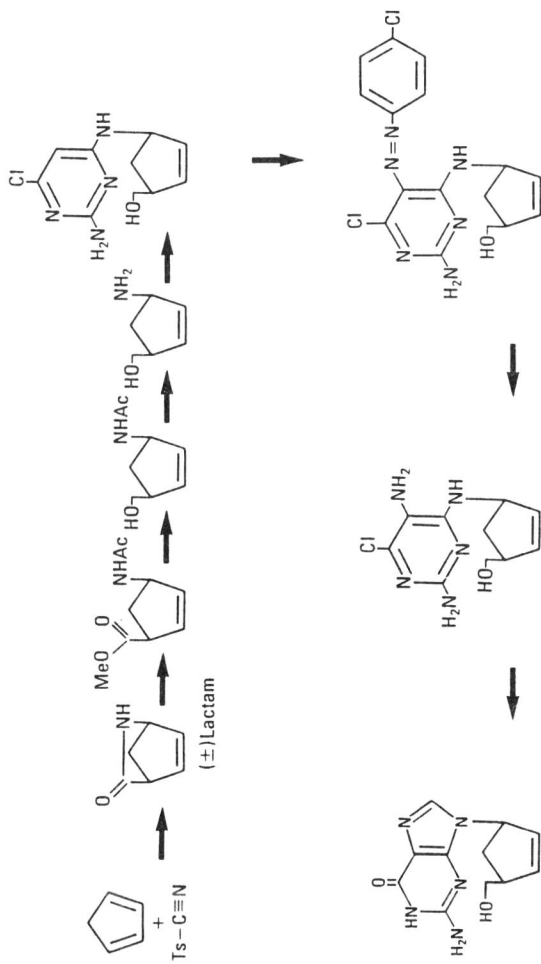

* J. Org. Chem. *43* (12) 1978 p.2311-2320
* J. Med. Chem. *33* (1) 1990 p.17-21

STRAIN SELECTION AND CHARACTERISATION

Figure 2 describes the stereochemical course of the hydrolysis of the racemic lactam (±)-2-Azabicyclo[2.2.1] hept-5-en-3-one catalysed by whole cell preparations of strains designated ENZA-1 and ENZA 20.

Both strains were isolated from the environment under conditions designed to select for organisms capable of growth on a range of N-acyl compounds as the sole source of carbon and energy.

Subsequent to isolation ENZA-1 and ENZA-20 were grown up separately in liquid culture for between 24 and 48 hours and the resulting cells were harvested by centrifugation.

The stereochemical course of the lactam hydrolase reaction catalysed by the respective cultures was characterised in the following manner:

Cells of each organism were incubated at 25°C in the presence of the lactam and the reduction in lactam concentration was followed by means of u.v. detection subsequent to a reverse phase H.P.L.C. separation. The reaction was quenched at 55% chemical conversion by harvesting the cells and the lactam remaining was isolated and characterised by optical rotation and chiral shift N.M.R. The results are summarised below:

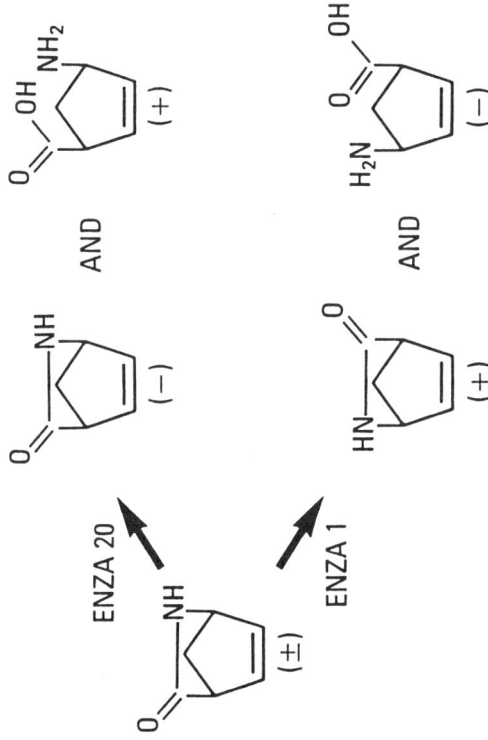

Figure 2

Whole cell mediated resolution of (±) γ-Lactam

ORGANISM	ROTATION OF LACTAM	PURITY BY CHIRAL SHIFT N.M.R.
ENZA-1	(+)	>98% e.e.
ENZA-20	(-)	>98% e.e

WHOLE CELL MEDIATED RESOLUTION OF

(±)-2-AZABICYCLO[2.2.1]HEPT-5-EN-3-ONE

Figure 3 describes typical reaction profiles for the hydrolysis of the racemic lactam by cultures of ENZA-1 and ENZA-20, in aqueous solution buffered to pH 7.0. In both the examples shown reaction composition and conditions were as follows:

[Biomass] $=$ 6g dry wgt l^{-1}

[±Lactam] $=$ 50gl^{-1}

[NaH$_2$PO$_4$] $=$ 50mM

Temp $=$ 25°C

Decrease in lactam concentration and coincident increase in amino acid concentration were followed by means of u.v. detection after separation by reverse phase HPLC. The cells were harvested at 55% chemical conversion and the lactam remaining was extracted into CH$_2$Cl$_2$, crystallised and characterised by Chiral Shift N.M.R.

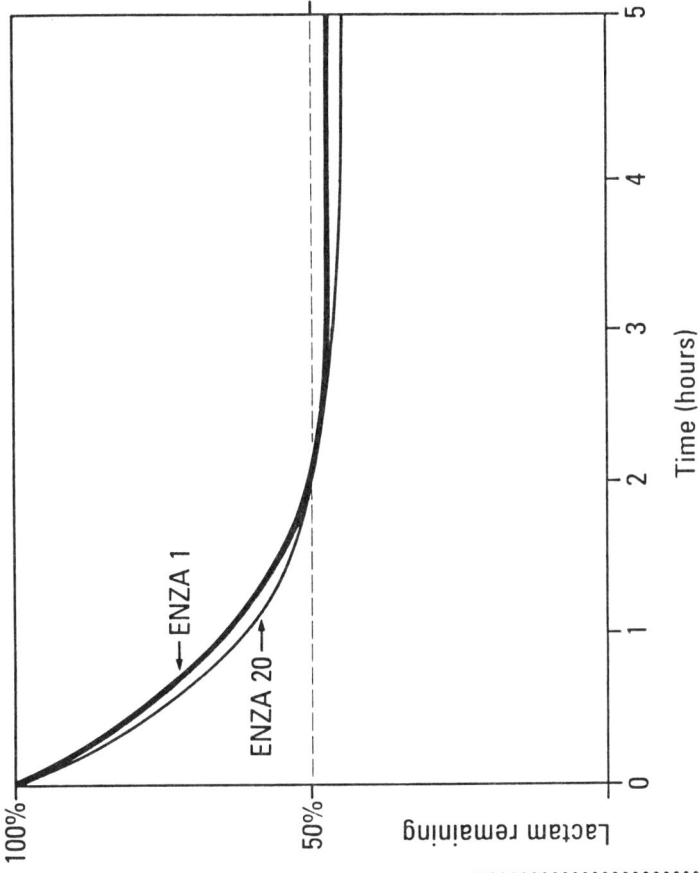

Figure 3

Kinetic
resolution of
($\pm$) γ-Lactam

OPTICAL PURITY OF LACTAM RECOVERED FROM

THE WHOLE CELL BIOTRANSFORMATION REACTION

Figure 4 reproduces spectra obtained as a result of characterisation of enantiomeric purity of the (-)-2-Azabicyclo
[2.2.1]hept-5-en-3-one by the technique of Chiral Shift
N.M.R. (^{1}H, 270MHz, $CDCl_3$). Spectrum 1 is a high resolution
proton spectrum of the racemic lactam (±)-2-
Azabicyclo[2.2.1.]hept-5-en-3-one demonstrating the following
resonances.

CHEMICAL SHIFT (δ)	INTEGRATION	SIGNAL	ASSIGNMENT
6.8	1H	MULTIPLET	VINYLIC
6.7	1H	MULTIPLET	VINYLIC
5.5 - 6.2	1H	BROAD	N - H
4.4	1H	MULTIPLET	METHINE
3.2	1H	MULTIPLET	METHINE
2.4	1H	MULTIPLET	METHYLENE
2.2	1H	MULTIPLET	METHYLENE

Spectrum 2 demonstrates the validity of the method employed
for assaying the enantiomeric excess by characterising the
effect of adding a Lanthanide Chiral Shift reagent (C.S.R) to
the racemic lactam:

115

Figure 4 Spectrum 1

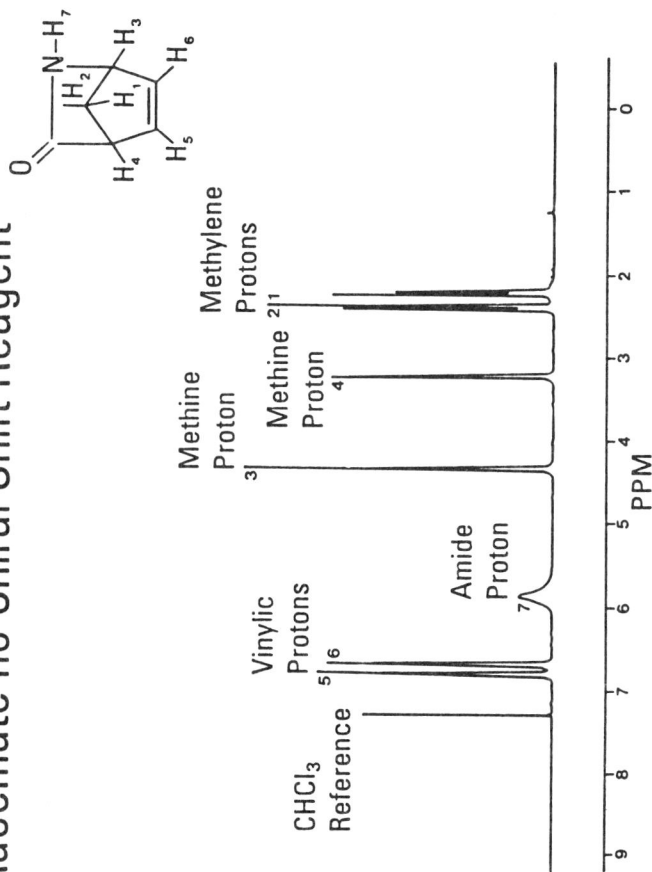

^1H NMR of Racemic Lactam and pure (−) Lactam in the presence of a Chiral Shift Reagent (CSR)

Racemate no Chiral Shift Reagent

Figure 4 Spectrum 2

¹H NMR of Racemate with Chiral Shift Reagent

¹H NMR of Racemic Lactam and pure (–) Lactam in the presence of a Chiral Shift Reagent (CSR)

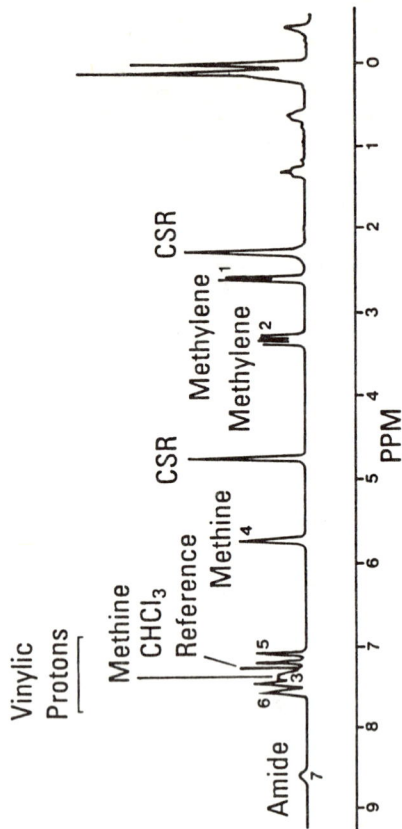

Figure 4 Spectrum 3

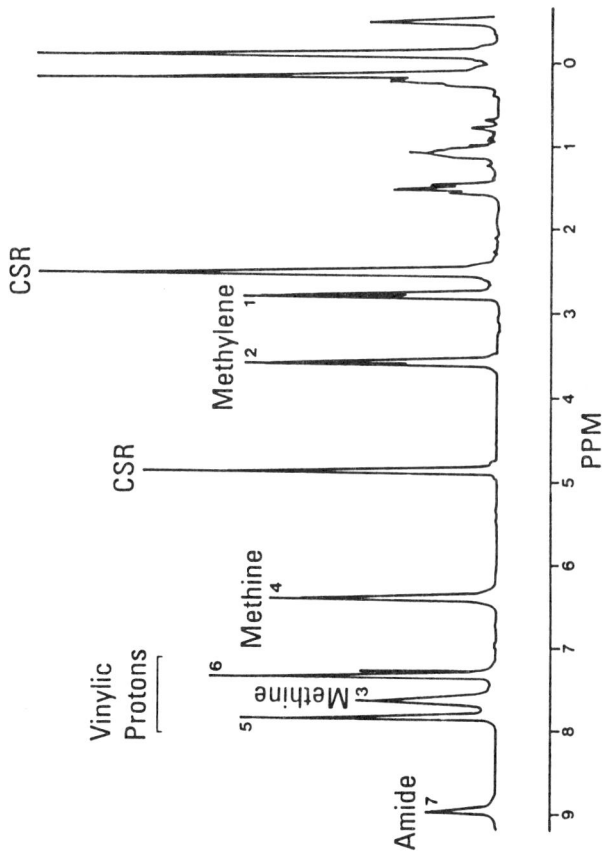

$^1H\ NMR$ of (–) Lactam with Chiral Shift Reagent

Racemic Lactam
and pure
(–) Lactam in the
presence of a
Chiral Shift
Reagent (CSR)

(a)   Both vinylic protons are shifted towards the $CHCl_3$ reference peak. Each vinylic multiplet is split to give signals of equal intensity, in accord with a 50:50 enantiomeric ratio.

(b)   A methine proton is shifted down field from 3.2 to give a poorly resolved broad double signal at δ7.2 -δ7.4.

(c)   One of the two methylene protons at δ 2.5 is resolved to a double doublet atδ3.3 -δ3.4.

Spectrum 3 results from addition of C.S.R to a sample of the (-) lactam. In this example both vinylic protons appear as single resonances (δ 7.8 and δ 7.3) as does the resonance for the methine proton (δ7.6). Also the most deshielded methylene proton (δ 3.6) has not been resolved into a double doublet as is the case for its counterpart in the racemic mixture (spectrum 2,δ3.3 -δ3.4).

These data are consistent with an enantiomeric purity of > 98% e.e. for the (-) lactam.

SUMMARY

Enzymatix has developed technology for the resolution of the lactam, ($\pm$)-2-Azabicylo[2.2.1]hept-5-en-3-one using two novel whole cell biocalysts. The use of such biocalysts allows rapid, facile, multi-kilo production of both chiral forms of the lactam at very high enantiomeric excess.

STEREO- AND REGIO-SPECIFIC ENZYMIC REACTIONS:
Chemical Exploitation Of The The Aromatic Dihydroxylation System

DAVID A. WIDDOWSON* AND DOUGLAS W. RIBBONS[†]

Departments of Chemistry* and Biotechnology[†], Imperial College of Science, Technology and Medicine; London SW7 2AY

ABSTRACT

The dihydroxylation of aromatic rings by mutants of *Pseudomonas putida*, blocked in the rearomatisation step, allows access to a series of *cis*-cyclohexadienediols with a variety of functionality around the ring. The reaction occurs with high regio- and stereo-specificity but low substrate specificity and is therefore very suitable for use in general organic synthesis. The scope of the reaction, the applications to complex molecule synthesis and the prospects for general use is described

INTRODUCTION

The ever increasing sophistication of organic synthesis in the chemical industry has led to a demand for more and more elaborate feedstocks and processes. There is a particular demand for materials or methods which will provide substances which are chirally pure or of high enantiomeric excess and to satisfy this demand, several options present themselves. These include:—

❑ recourse to the 'chiral pool' of readily available naturally derived simple materials either as starting materials or as resolving agents.

❑ biosynthesis of the target substance or a structurally very similar species *via* fermentation.

❑ generation of the asymmetry and perhaps much of the functionality by one or more 'biotransformation' processes.

Biosynthesis, *via* microbial fermentation, of the complete target molecule or a close analogue is clearly a very valuable access to complex molecules and indeed its use predates chemical synthesis, but it is limited in scope in terms of the total number of desirable targets.

Of much greater applicability is the use of microbes as 'synthetic reagents'. Although this usage is not new, the ready and increasingly rapid availability of mutants or genetically engineered microbes means that microbial 'reagents', tailored to a particular transformation are a real option to be considered in synthesis design. Once developed, the mutants show some remarkable advantages over conventional chemical reagents and over isolated enzymes. Thus they have the ability, in favourable cases, to:—

❑ be self-sustaining on a simple medium.
❑ bring about unit transformations upon a wide range of substrates.
❑ avoid over reaction by the use of the mutated/engineered blocks in the metabolic pathway.
❑ accept unnatural substrates and therefore be important in general synthesis.
❑ operate usually with total regio- and stereo-specificity.
❑ bring about transformations which cannot be achieved by organic synthesis.

These transformations will of course be limited to those normally occurring in the organism but in principal, any microbe can be manipulated and any non-vital process, known to occur in microbes, could be accessed by mutation to block the subsequent reaction although other factors may limit the feasibility of the process. The general applications of biotransformations in isolated enzymes and whole organisms have been reviewed [1].

The remainder of this account will illustrate these principles by concentrating on one transformation, aromatic dihydroxylation, in one organism, *Pseudomonas putida*.

AROMATIC DIHYDROXYLATION

Within the general theme of regio- and stereo-specific biotransformations for organic synthesis, the production of cis-dihydroxycyclohexadienes (such as **1**) have a particular attraction. This structural array would be difficult to attain by conventional synthesis and yet the system contains the synthetically useful features of stereodefined multifunctionality. The dihydroxylation of aromatic rings (e.g. Scheme 1, the *p*-cymene pathway) by micro-organisms has been known for many years [2] and the production of mutant strains which are blocked in the rearomatisation step were first described in 1968 [3]. Many of these have now been produced and some developed to a commercial level [4] making the corresponding cyclohexadienediols (as 1) are readily available.

Scheme 1

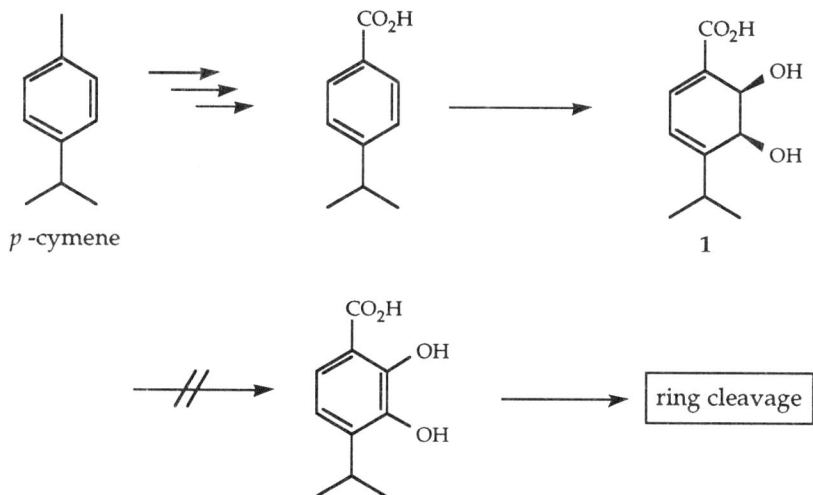

Clearly defined minimum structural requirements have been established for the substrates of a number of the mutants [2]. This site specificity for a minimal requirement is high yet the substrate specificity can be low with respect to further ring substitution. This is a important for the general usefulness of the mutants as too high a specificity for the whole substrate would result in a need for an excessive number of mutants for general synthetic application.

Many differently functionalised diols are therefore available from the different mutants (Scheme 2) and the absolute stereochemistry of several of these have been established [5] to be as shown in Scheme 2. Yields of up to 30g/l of broth can be obtained in favourable cases [6] although ≈ 1 – 10 g/l are more common. Few of these transformations have yet been exploited to any extent in organic synthesis and most use has been made of the simple 'benzene-*cis*-diol' (*cis*-5,6-dihydroxycyclohexa-1,3-diene) (2, R = H). The functional groups R and R' (in Scheme 2) show considerable diversity [7] and remarkable tolerance by the enzymes concerned. Dependant upon the mutant and transformation concerned, they have been shown to include F, Cl, Br, I, Me, Et, n-Pr, i-Pr, n-Bu, t-Bu, CF_3, vinyl, phenyl, allyl, benzyl, EtO, n-PrO, n-BuO and PhCO and also the fused ring naphthalene and tetralin systems.

Scheme 2

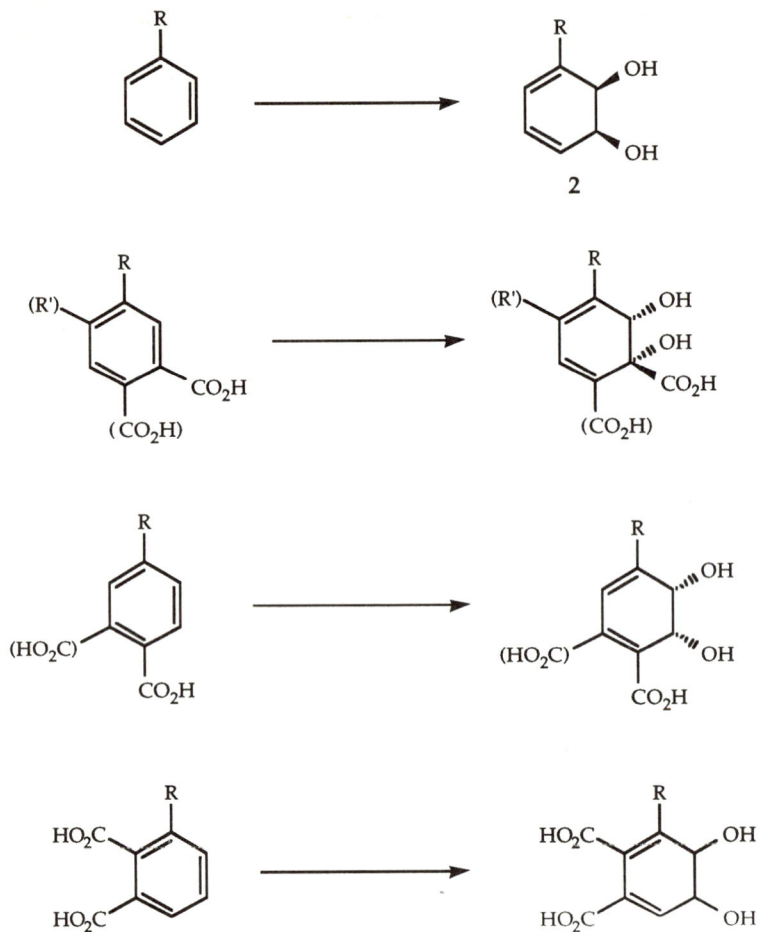

The manner in which the various structural features of the dihydroxydienes have been used in the synthesis of a range of synthetic intermediates is illustrated below. Thus far the applications do not appear to have reached a commercial level, despite the obvious potential.

SYNTHETIC APPLICATIONS

1. (+)Pinitol (3).

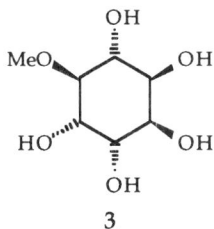

3

This cyclitol has been shown to act as a feeding stimulant and an inhibitor of larval growth in *Eurema hecabe mandarina* [8] and *Heliothis zea* [9] respectively. Hypoglycemic and antidiabetic activity has been detected in mice [10]. The synthesis of the compound, by Ley and Sternfeld [11], started from the achiral 'benzene-*cis*-glycol' and therefore could not use the full potential of its biological source for stereoselectivity and had to incorporate a resolution step. Nevertheless the biotransformation allowed a short efficient the synthetic sequence (Scheme 3) to be developed.

Scheme 3

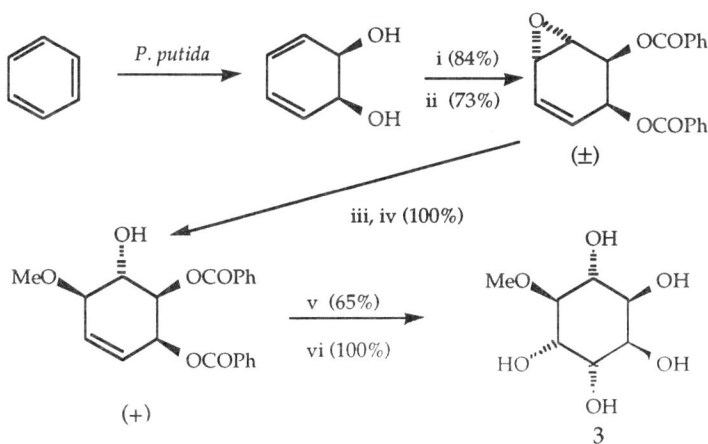

i, PhCOCl/DMAP-py; ii. mCPBA; iii, resolution *via* menthoxyacetate ester; iv, MeOH/camphorsulphonic acid; v, OsO$_4$, N-methylmorpholine-n-oxide; vi, Et$_3$N/MeOH/H$_2$O.

Starting from the same diol, but in a more complex sequence, the same authors have reported [12] the synthesis of the hormonal secondary messenger,

IP$_3$ (4).

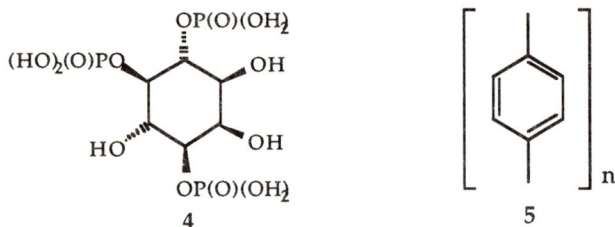

2.Polyphenylene (5).

An interesting use of 'benzene-*cis*-glycol' which does not seem to have been developed is the synthesis of polyphenylene [13], a polymer which was expected to have interesting electrical properties when suitably doped. It was straightforwardly prepared by polymerising a diester of 'benzene-*cis*-glycol' followed by pyrolytic elimination of the two ester groups to generate the aromatic rings (Scheme 4).

Scheme 4

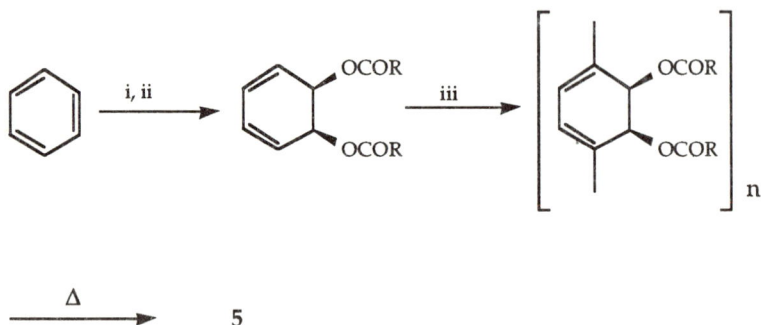

i, *P. putida* ; ii, (RCOCl/base); iii, radical initiator.
R = Me, Ph, or OMe

3. Prostanoid (6) and Terpenoid (7) Synthons.

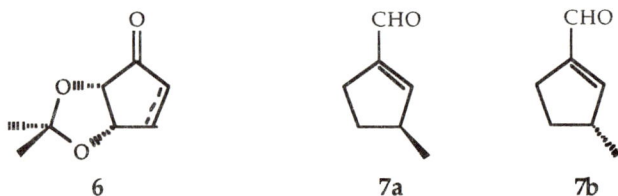

The synthesis by Hudlicky *et al.* of the prostanoid intermediate (6) develops the potential of the metabolites more fully by starting with the homochiral diol (2, R = Me) derived from toluene [14]. Controlled cleavage of the diene by ozonolysis or with singlet oxygen and retention of the diol chirality opened the way to a simple sequence to the chiral synthon (6) (Scheme 5).

Scheme 5

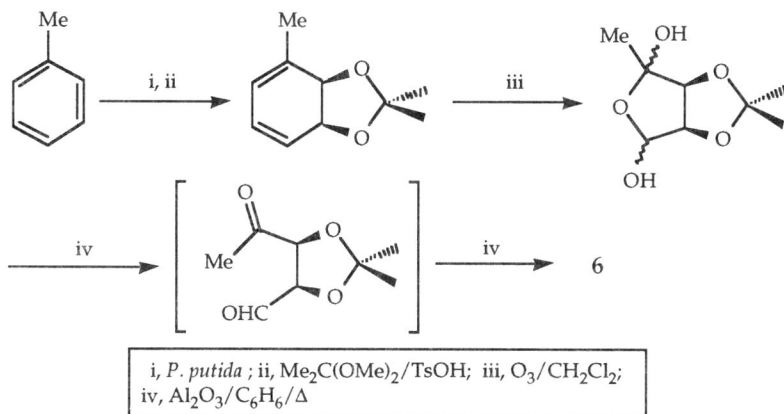

i, *P. putida* ; ii, Me$_2$C(OMe)$_2$/TsOH; iii, O$_3$/CH$_2$Cl$_2$; iv, Al$_2$O$_3$/C$_6$H$_6$/Δ

The synthesis of the terpenoid synthons (7) (Scheme 6) made less effective use of the biotransformation product [14]. The diene was not required and the catalytically hydrogenation to the saturated cyclohexane ring proved to be non-stereospecific. The resulting diastereomeric diols had to be separated and each taken on to the enals (7). The redeeming feature of this was the obtention of both enantiomers without a separate resolution step.

Scheme 6

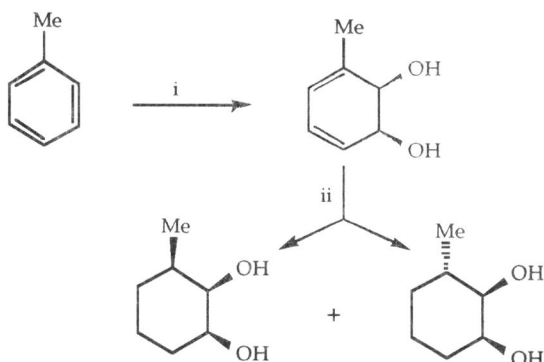

continued over

Scheme 6 (continued)

i, *P. putida* ; ii, H_2/Pd-C; iii, HIO_4; iv, AcOH/piperidine

4. (–)-Zeylena (8).

8

This is the structurally the most complex product yet made from a dienediol (2). The synthesis [15] uses all of the positive features of the biotransformation with the diene and diol functionalities and the chirality all developed into the target molecule and a biotransformation with a sensitive substrate, styrene. The reactive vinyl group in the substrate comes through the hydroxylation process unscathed. Because of the high degree of functionality built into the biotransformation product, the synthetic route is short for such a complex structure, at 10 steps beyond the biotransformation. The full Scheme is given below.

Scheme 7

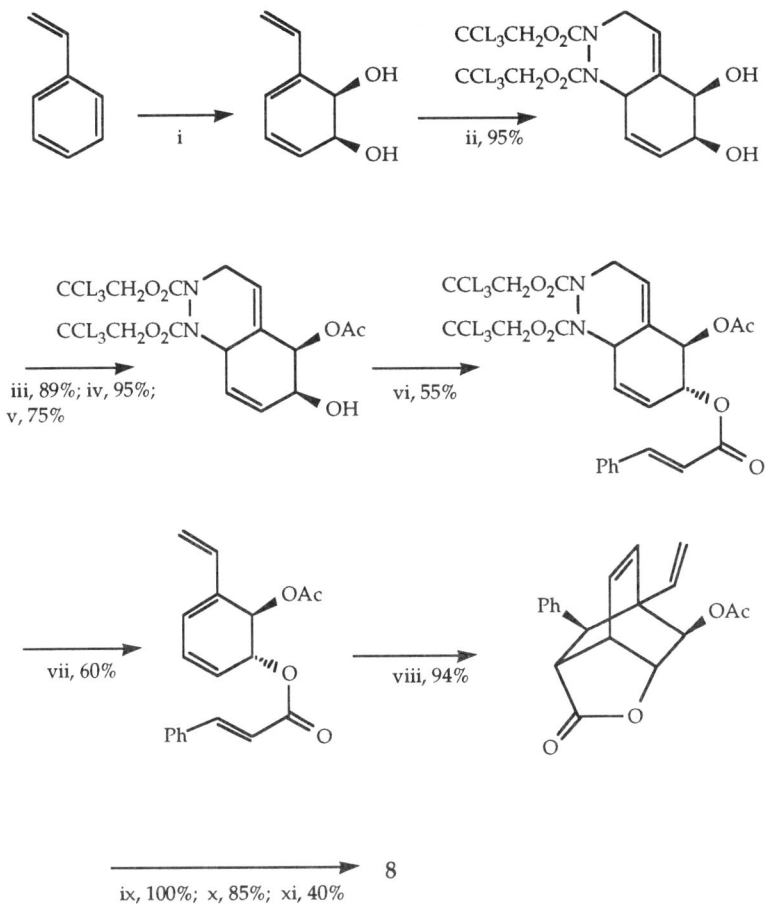

i, *P. putida* ; ii, bis(2,2-trichloroethyl)azodicarboxylate; iii, R_3SiCl/imidazole/ DMF/4°C; iv, Ac_2O/py/DMAP; v, n-Bu_4NF/THF/rt; vi, Ph_3P/$(EtO_2CN)_2$/ $PhCH=CHCO_2H$; vii, Zn(Cu); viii, C_6H_6/110°C; ix, O_3/CH_2Cl_2/Me_2S; x, $NaBH_4$; xi, $PhCO_2H$/i-BuOCOCl/Et_3N/THF.

5. Organoiron Complexes (9).

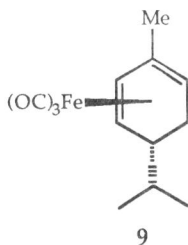

9

Attachment of a transition metal to the diene moiety adds an additional dimension to the synthetic potential of the dienediols. Transition metal complexes are undergoing a rapid development as reagents for organic synthesis [16] and offer unique reactivities frequently under exceptionally mild conditions and with high stereocontrol. A major problem in conventional approaches to these complexes is the limited access to homochiral material. The microbially derived dienediols resolve this problem for diene ligands and at the same time attachment of the metal can stabilise the inherently labile ligand. Thus the combination of the two chemistries offers considerable scope for synthetic development.

Progress has been reported [17] with the irontricarbonyl complexes (Scheme 8) in synthesising intermediates for chiral alkaloid synthesis.

Scheme 8

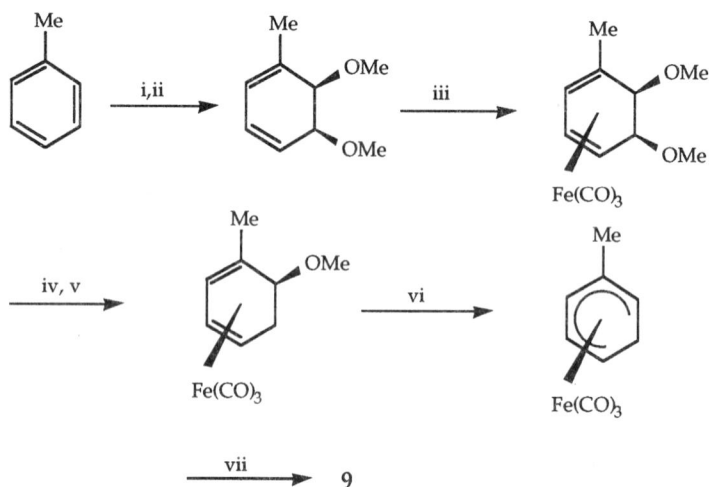

i, *P. putida* ; ii, MeI/KOH; iii, Fe$_2$(CO)$_9$; iv, Ph$_3$C$^+$ BF$_4^-$ / NH$_4$PF$_6$; v, NaBH$_4$; vi, CF$_3$CO$_2$H/NH$_4$PF$_6$; vii, NaCH(CO$_2$Me)$_2$.

CONCLUSIONS

The above syntheses illustrate some of the features which a biotransformation can bring to organic synthesis. The general attributes which make their use so attractive are the access they provide to:—

❑ optically pure compounds.
❑ chemically labile products.

❏ novel transformations.

❏ regio-controlled reactions.

❏ chemospecific transformations.

❏ substrate specificity; high specificity for a part of a substrate molecule and low specificity for the remainder.

❏ new mutants with slightly differing specificities and thereby increasing the spectrum of substrates

Biotransformations can provide a ready source of many chiral starting materials, as above, but an additional feature of the broad substrate specificity is the potential for application at a much later stage in a synthetic sequence such that generation of the sensitive dienediol functionality can be delayed to a more appropriate point.

The aromatic dihydroxylation alone has considerable and as yet largely unrealised potential. Already the use of yeasts in stereo- and regio-controlled redox reactions is widespread. The whole gamut of biotransformations is set to make a major impact on organic synthesis at research and commercial level.

REFERENCES

1. J.B. Jones, *Tetrahedron*, 1986, **42**, 3351, and the many references cited there.

2. J.J. De Frank and D.W. Ribbons, *J. Bacteriol.*, 1977, **129**, 1356, *ibid.* 1365.

3. D.T. Gibson, J.R. Koch and R.E. Kallio, *Biochemistry*, 1968, 7, 2653, *ibid.* 3795.

4. S.C. Taylor and S. Brown, *Performance Chem.*, 1986, (Nov.) 20.

5. S.J.C. Taylor, D.W. Ribbons, A.M.Z. Slawin, D.A. Widdowson and D.J. Williams. *Tetrahedron Lett.*, 1987, 28, 6391; V.M. Kobal, D.T. Gibson, R.E. Davis and A. Garza, *J. Am. Chem. Soc.*, 1973, 95, 4420.

6. D.W. Ribbons, S.J.C. Taylor and D.A. Widdowson, unpublished results.

7. D.W. Ribbons, C.T. Evans, J.T. Rossiter, S.J.C. Taylor, S.D. Thomas, D.A. Widdowson and D. J. Williams, NATO Conference, Lisbon, 1989, in press.

8. A. Namata, K. Hokimoto, A. Shimada, H. Yamaguchi and K. Takaishi, *Chem. Pharm. Bull.*, 1979, **27**, 602.

9. J.C. Reece, B.G. Chan and A.C. Waiss Jr, *J. Chem. Ecol.*, 1982, **8**, 1429; D.L. Drewer, R.G. Binder, B.G. Chan, A.S. Waiss Jr, E.E. Hartwig and G.L. Beland, *Experientia*, 1979, **35**, 1182.

10. C.R. Narayan, D.D. Joshi, A.M. Miyumdar and V.V. Dhekne, *Curr. Sci.*, 1987, **56**, 139.

11. S.V. Ley and F. Sternfeld, *Tetrahedron Lett.*, 1988, **29**, 5305.

12. S.V. Ley and F. Sternfeld, *Tetrahedron*, 1989, **45**, 3463.

13. D.G.H. Ballard, A. Courtis, I.M. Shirley and S.C. Taylor, *J. Chem. Soc., Chem. Commun.*, 1983, 954.

14. T. Hudlicky, H. Luna, G. Barbieri and L.D. Kwart, *J. Am. Chem. Soc.*, 1988, **110**, 4735.

15. T. Hudlicky, G. Seone and T. Pettus, *J. Org. Chem.*, 1989, **54**, 4239.

16. For a general account see, J.P. Collman, L.S. Hegedus, J.R. Norton and R.G. Finke, 'Principles and Applications of Organotransition Metal Chemistry', University Science Books / Oxford University Press, Mill Valley, California, 2nd edn., 1987.

17. P.W. Howard, G.R. Stephenson and S.C.Taylor, *J. Chem. Soc., Chem. Commun.*, 1988, 1603.

ASYMMETRIC SYNTHESIS OF COMPLEX OLIGOSACCHARIDES

KURT G.I. NILSSON
Carbohydrates International AB, Arlöv, Sweden
Present Address: Chemical Center, University of Lund,
P.O.Box 124, S-221 00 Lund, Sweden

ABSTRACT

A number of disaccharide sequences common in glycoconjugates
have been prepared employing glycosidases and appropriate
donor and acceptor glycosides.
Sialylated trisaccharides were prepared by the sequential use
of glycosidase and glycosyltransferase. Hexosamine-containing
disaccharides of the type HexNAc-Gal (e.g. GalNAcα1-3Gal,
GlcNAcβ1-3Gal) were formed from suitable donor and acceptor
glycosides, employing glycosidases from Chamelea gallina.
Moreover, an α-L-fucosidase from the same mollusc catalysed
the efficient synthesis of Fucα1-6Galβ-OMe.

INTRODUCTION

The carbohydrate structures of glycoproteins and glycolipids
are involved in a variety of biological processes, such as
intracellular migration and secretion of glycoproteins, cell-
cell interaction, oncogenesis, interaction of cell-surfaces
with pathogens, etc (1-3). Shorter fragments (di- or trisaccharides) of these complex carbohydrate chains are often
sufficient for full biological activity (e.g. blood group
determinants, tumor-associated antigens, receptors for
pathogens).

The availability of such oligosaccharides of well-defined
structure and in sufficient quantity is critical for
applications and fundamental studies. Isolation from natural
sources is a complex task and is not economical on a larger

scale due to the low concentration of the structures of interest present in the complex mixtures of carbohydrates obtained from biological sources. Selective chemical synthesis requires many protection and deprotection steps, which result in low yields. Moreover, stereospecific chemical synthesis of oligosaccharides, especially of the important α-sialylated structures, is difficult (4). Therefore, biocatalysts are attractive as they should allow the stereospecific and regioselective synthesis with a minimum of reaction and purification steps.

Facile methods for the synthesis of various oligo- saccharides based on glycosidases, synthases and glycosyl- transferases have been developed (5) and are briefly summarised in this paper. In addition, the formation of various disaccharides present in antigenic determinants and carbohydrate receptors for bacteria, of the type HexNAc-Gal and Fuc-Gal, employing glycosidases from Chamelea gallina is described.

MATERIALS AND METHODS

The general and instrumental methods were as described (6-8). The structure of isolated products was determined with NMR (^{13}C- and ^{1}H-NMR; Varian XL 200 instrument) as reported (6). All reagents employed were of analytical grade and used as supplied. The monosaccharide glycosides were obtained from Sigma (St. Louis, Missouri). All sugars were of the D-configuration (except the L-fucosylpyranosides). The reactions were followed by TLC, HPLC and spectrophotometric measurement of liberated nitrophenol. The glycosidase preparation from Chamelea gallina was obtained by homogenisation of the liver, extraction with distilled water and fractionating the supernatant with ammonium sulphate as described (9).

Synthesis of HexNAc-Gal Glycosides with Glycosidases from Chamelea gallina

In a typical experiment, the ammonium sulphate precipitate (100 mg) was added to sodium phosphate buffer (20 ml, 0.04 M, pH 5.8) containing p-nitrophenyl 2-acetamido-2-deoxy-ß-D-galactopyranoside (GalNAcß-OPhNO$_2$-p; 75 mM) and methyl ß-D-galactopyranoside (Galß-OMe; 0.6 M) and the mixture was stored with gentle agitation for five days at 37 °C. The products were isolated by gel chromatography (Sephadex G10) and semipreparative HPLC (Lichrosorb NH$_2$-silica; 70 % aqueous acetonitrile). Crystallisation from ethanol gave GalNAcß1-3Galß-OMe (36 mg) which was pure according to NMR (^{13}C; ^{1}H), HPLC and elemental analysis.

The synthesis and isolation of GlcNAcß-Gal- and GalNAcα1-3Gal glycosides were achieved analogously using the substrate glycosides shown in Table 2.

Synthesis of Fucα1-6Galα-OMe with α-L-Fucosidase from Chamelea gallina

Ammonium sulphate precipitate (0.5 g) from Chamelea gallina was added to a mixture of p-nitrophenyl α-L-fucopyranoside (0.4 g) and methyl ß-D-galactopyranoside (2.0 g) in sodium acetate buffer (20 ml, 0.05 M, pH 5.5). After storage for 70 h at 37 °C, the mixture was subjected to column chromatography (Sephadex G10). The fractions containing disaccharide were acetylated (pyridine-acetic anhydride) and column chromato-graphy on Kieselgel 60 (toluene-ethyl acetate, 1:1) gave pure acetylated Fucα1-6Galß-OMe (280 mg). The structure was confirmed by NMR (characteristic downfield shifts of the C-6, H-2, H-3, and H-4 resonances). Deacetylation with methanolic sodium methoxide gave the title disaccharide.

RESULTS AND DISCUSSION

Glycosidases catalyse the stereospecific synthesis of oligo-saccharides in equilibrium or transglycosylation reactions (5). The stereospecificity and high availability of glycosidases make them highly attractive for glycoside synthesis. The low regioselectivity and preponderant formation of undesired 1-6-linkages have hampered their wider application for the synthesis of biologically active oligosaccharides. However, it has been shown that the regioselectivity of the trans-glycosylation reactions is influenced by a number of parameters, such as reaction temperature (10), concentration of organic cosolvent, the reactivity of donor (11), the anomeric configuration of the acceptor glycoside and the nature of its aglycon (6,8). Especially the use of various acceptor glycosides ($HOAR_2$ in the annexed scheme) represents a powerful strategy to change the regioselectivity in the desired direction. A schematic outline of the reactions involved is given below:

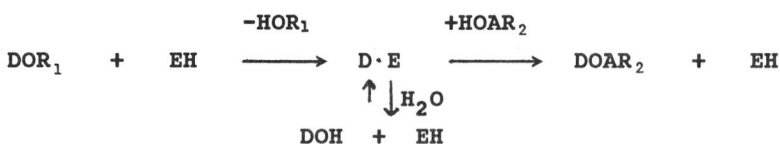

$$DOR_1 \ + \ EH \ \xrightarrow{\ \ -HOR_1\ \ } \ D \cdot E \ \xrightarrow{\ \ +HOAR_2\ \ } \ DOAR_2 \ + \ EH$$

$$\uparrow \downarrow H_2O$$

$$DOH \ + \ EH$$

(DOR_1 = donor glycoside, EH = glycosidase, $DOAR_2$ = product)

Some results obtained with this method are summarised in Table 1.

The yields in these reactions were in the range 10-45 % depending on enzyme and reaction conditions (6-8,10,11). For example, higher yields of products were obtained with methyl glycosides as acceptors than with the corresponding nitro-phenyl glycosides due to the lower solubility of the latter.

The regioselectivity was in the range 70-95 %, except the formation of ß1-3 and ß1-4-linked Gal-GlcNAc-OMe glycosides which were formed in a ratio of 55:50. Since the reactions were stereospecific and no anomerisation occured, purification of the products by column chromatography was straightforward.

In addition, glycosides that can be utilised variously were obtained. For example, allyl, benzyl, thioethyl and tri-methylsilylethyl glycosides are useful for temporary anomeric protection and nitrophenyl glycosides can be used for the preparation of neoglycoproteins after chemical modification.

TABLE 1
Glycosidase-catalysed synthesis of disaccharide glycosides.

Glycosyl donor	Acceptor	Main glycosides formed
α-Galactosidase		
Gal α-OPhNO$_2$-p	Gal α-OMe	Gal α1-3Gal α-OMe
Gal α-OPhNO$_2$-o	Gal α-OPhNO$_2$-o	Gal α1-2Gal α-OPhNO$_2$-o
Gal α-OPhNO$_2$-p	Gal α-OPhNO$_2$-p	Gal α1-3Gal α-OPhNO$_2$-p
Gal α-OPhNO$_2$-p	Gal α-OCH$_2$CH=CH$_2$	Gal α1-3Gal α-OCH$_2$CH=CH$_2$
Gal α-OPhNO$_2$-p	Galß-OMe	Gal α1-6Galß-OMe
Gal α-OPhNO$_2$-p	GalNAc α-OEt	Gal α1-3GalNAc α-OEt
ß-Galactosidase		
Galß-OPhNO$_2$-p	Gal α-OMe	Galß1-6Gal α-OMe
Galß-OPhNO$_2$-p	Galß-OMe	Galß1-3Galß-OMe
Lactose	Allylalcohol	Galß1-OCH$_2$CH=CH$_2$
		Galß1-3Galß-OCH$_2$CH=CH$_2$
Lactose	Benzyl-alcohol	Galß1-OCH$_2$Ph
		Galß1-3Galß-OCH$_2$Ph
Lactose	Trimethyl-silylethanol	Galß1-OEtSiMe$_3$
		Galß1-3Galß-OEtSiMe$_3$
Galß-OPhNO$_2$-p	GalNAc α-OEt	Galß1-3GalNac α-OEt
Galß-OPhNO$_2$-p	GlcNAcß-OMe	Galß1-3GlcNAcß-OMe
		Galß1-4GlcNAcß-OMe
Galß-OPhNO$_2$-o	GlcNAcß-OEtSiMe$_3$	Galß1-3GlcNAcß-OEtSiMe$_3$

As seen from Table 1, monosaccharide glycosides can be prepared _in situ_ and this is exemplified by the synthesis of allyl, benzyl and trimethylsilylethyl glycosides from lactose and alcohols (7).

The disaccharide sequences shown in Table 1 occur widely in Nature. For example, Gal α1-3Gal is a constituent of blood group B, Galß1-3GlcNAc is present in Lewis antigenic determinants, Gal α1-3GalNAc has been isolated from terato-carcinoma cells and Galß1-3GalNAc (the T-antigenic determinant) is present on the cell surfaces of human carcinomas (2).

Synthesis of Trisaccharides
There are some examples of glycosidase-catalysed synthesis of trisaccharide sequences present in glycoconjugates. This was achieved by either using the same glycosidase in two consecutive transglycosidation reactions, e.g. synthesis of

Galß1-3Galß1-4GlcNAcß-OEtSiMe$_3$ from Galß-OPhNO$_2$-o and GlcNAcß-OEtSiMe$_3$ (8), or by using two glycosidases in sequence, e.g. synthesis of Galα1-3Galß1-4GlcNAc with ß-galactosidase and α-galactosidase (12). Although this approach is simple, yields are often relatively low.

In an alternative approach, glycosidase and glycosyl-transferase were used in sequence for the synthesis of sialylated trisaccharides (8). High final yields were obtained and the regiospecificity of the transferase minimised the amount of purification steps required. The method is exemplified below:

1. Galß-OPhNO$_2$-o + GalNAcß-OEtBr ⟶ Galß1-3GalNAcß-OEtBr

CMP-Neu5Ac
2. Galß1-3GalNAcß-OEtBr ⟶ Neu5Acα2-3Galß1-3GalNAcß-OEtBr

In the first reaction, ß-galactosidase from bovine testes was used and in the second reaction ß-D-galactoside α2-3sialyl-transferase from porcine submaxillary glands was employed. The donor in the second reaction, CMP-Neu5Ac, was easily prepared from CTP and Neu5Ac using a crude CMP-sialate synthase preparation (13). The sialyltransferase and the CMP-sialate synthase were immobilised to tresyl chloride activated agarose to allow their reutilisation and simplify purification of products. The immobilised sialyltransferase and synthase were reused three times without noticeable decrease in activity. High yields of CMP-Neu5Ac (91 %) and of sialylated trisaccharide (80 %) were obtained (8,13).

Synthesis with Glycosidases from <u>Chamelea gallina</u>
Disaccharide structures of the type HexNAc-Gal are found in a number of important antigenic determinants and some are receptor structures for bacteria. Thus, GalNAcα1-3Gal is part of blood group A (GalNAcα1-3(Fucα1-2)Gal), GalNAcß1-3Gal is terminal unit of blood group P, and GlcNAcß1-3Gal acts as a receptor for <u>Streptococcus pneumoniae</u> (3) and is a constituent of tumor-associated antigens as is GalNAcß1-3Gal (2). It was

found that a crude preparation of glycosidases from the liver
of the mollusc Chamelea gallina could be used for the
preparation of these and some additional structures (Table 2).

TABLE 2
Synthesis of disaccharide glycosides with a crude glycosidase
preparation obtained from Chamelea gallina.

Donor	Acceptor	Main Product
GlcNAcß-OPhNO$_2$-p	Manα-OMe	GlcNAcß1-6Manα-OMe
GlcNAcß-OPhNO$_2$-p	Galα-OMe	GlcNAcß1-6Galα-OMe
GlcNAcß-OPhNO$_2$-p	Galß-OMe	GlcNAcß1-3Galß-OMe
		GlcNAcß1-6Galß-OMe
GalNAcß-OPhNO$_2$-p	Galß-OMe	GalNAcß1-3Galß-OMe
GalNAcα-OPh	Galα-OMe	GalNAcα1-3Galα-OMe
Fucα-OPhNO$_2$-p	Galß-OMe	Fucα1-6Galß-OMe

The yields in the above reactions were about 10% or lower, but
the yields could be increased at least threefold by using a
higher acceptor concentration. The simplicity of the reactions
and the possibility to use the same, crude enzyme preparation
for selective synthesis of several structures of interest make
the method attractive.

As seen from Table 2, the regioselectivity of the
hexosaminidase-catalysed reaction was influenced by the
anomeric configuration of the acceptor glycoside. With the
ß-linked methyl galactoside as acceptor, GlcNAcß1-3Galß-OMe
was obtained while the α-glycoside gave almost exlusive
formation of the ß1-6-linked isomer. This latter structure
and GlcNAcß1-6Man, are found in O- and N-glycoproteins
respectively (1). Increased GlcNAcß1-6Man branching has
been associated with metastasis (2).

CONCLUSION

Enzymatic methods that allow stereospecific and regioselective
synthesis of biologically active oligosaccharides and their
glycosides with a minimum of reaction steps are available.
Future developments will include possible ways to increase the

yields of glycosidase-catalysed transglycosylations (which
often are below 50 % today), application of glycosidases for
synthesis of higher oligosaccharides and production of
glycosyltransferases and their donor glycosides on a larger
scale.

REFERENCES

1. Biology of Carbohydrates, eds. V. Ginsburg and P.W.
    Robbins, Wiley, New York, 1984.

2. Hakomori, S., Tumor-associated carbohydrate antigens, Ann.
    Rev. Immunol., 1984, 2, 103-126.

3. Karlsson, K.-A., Animal glycosphingolipids as membrane
    attachment sites for bacteria, Ann. Rev. Biochem., 1989,
    58, 309-350.

4. Paulsen, H., Synthesis of complex oligosaccharide chains
    of oligosaccharides, Chem. Soc. Rev., 1984, 13, 15-45.

5. Nilsson, K.G.I., Enzymatic synthesis of oligosaccharides,
    Trends in Biotechnol., 1988, 6, 256-264.

6. Nilsson, K.G.I., A simple strategy for changing the regio-
    selectivity of glycosidase-catalysed formation of di-
    saccharides, Carbohydr. Res., 1987, 167, 95-103.

7. Nilsson, K.G.I., A simple strategy for changing the regio-
    selectivity of glycosidase-catalysed formation of di-
    saccharides: Part II, Enzymic synthesis in situ of various
    acceptor glycosides, Carbohydr. Res., 1988, 180, 53-59.

8. Nilsson, K.G.I., Enzymic synthesis of di- and tri-
    saccharide glycosides, using glycosidases and ß-D-
    galactoside 3α-sialyltransferase, Carbohydr. Res., 1989,
    188, 9-17.

9. Reglero, A. and Cabezas, J.A., Glycosidases of molluscs.
    Purification and properties of α-L-fucosidase from
    Chamelea gallina L., Eur. J. Biochem., 1976, 66, 379-387.

10. Nilsson, K.G.I., Influence of various parameters on the
    yield and regioselectivity of glycosidase-catalysed
    formation of oligosaccharide glycosides, Ann. N. Y. Acad.
    Sci., 1988, 542, 383-389.

11. Nilsson, K.G.I., A comparison of the enzyme-catalysed
    formation of peptides and oligosaccharides in various
    hydroorganic solutions using the nonequilibrium approach,
    Studies in Organic Chemistry, 1987, 29, 369-374.

12. Nilsson, K.G.I., Enzymatic synthesis of trisaccharides, Abstract, Eurocarb V, Fifth European Symposium on Carbo- hydrates, Prague, 1989, pp. C-76.

13. Nilsson, K.G.I., and Gudmundsson, B.-M., Synthesis of CMP- Neu5Ac and Neu5Acα2-3Galβ1-3GalNAcα-OEt with immobilised cytidin-5-monophosphosialate synthase and β-D-galactoside α2-3sialyltransferase, _Sialic Acids 1988,_ eds. R. Schauer and T. Yamakawa, Kieler Verlag Wissenschaft und Bildung, 1988, pp. 28-31.

USEFUL REDUCTION AND OXIDATION REACTIONS CATALYSED BY ENZYMES

STANLEY M. ROBERTS
Department of Chemistry,
University of Exeter,
Stocker Road, Exeter, Devon EX4 4QD, U.K.

1.   INTRODUCTION

The use of enzymes to catalyze interesting reduction and oxidation reactions has increased exponentially over the past ten years.   This article will illustrate some of the highlights from this work;   a more comprehensive review of the literature (up to late 1987) is available.[1] Some of the most useful reduction reactions that have been described involve the transformation of ketones into chiral secondary alcohols (Section 2.1) and the reduction of the carbon-carbon double bond in αβ-unsaturated carbonyl compounds to give useful optically active products (Section 2.2). In the field of enzyme-controlled oxidation reactions the oxidation of alcohols to aldehydes and ketones has received some attention (Section 3.1); more recently the regioselective oxidation of alicyclic compounds (Section 3.2) and the regio- and stereo-selective oxidation of aromatic compounds has attracted interest (Section 3.3).   Other noteworthy enzyme-catalysed oxidation reactions are described in Section 3.4.

2.   ENZYME-CATALYSED REDUCTION REACTIONS

2.1 Preparation of Secondary Alcohols

The synthesis of secondary alcohols from ketones can be accomplished using isolated partially purified enzymes or by utilizing whole-cell systems.   When using the partially purified protein, re-charging the enzyme with co-factor (NADH or NADPH) is necessary and co-factor recycling needs to be arranged.   The "pros and cons" of using an isolated enzyme or a whole-cell system for the reduction of carbonyl compounds has been discussed elsewhere.[2]

Simple ketones can be reduced to optically active secondary alcohols often with good to excellent optical purity of the products. Thus the ketone (1) has been reduced to the alcohol (2) (97% e.e.) using yeast and this alcohol was used to prepare one component of a pheromone.[3] Reduction of the ketone (3) with the enzyme *Thermoanaerobium brockii* dehydrogenase produces the natural product sulcatol (4) [50% yield (reaction terminated); 99% e.e.]. The dehydrogenase was re-constituted with the appropriate co-factor (NADPH) and the co-factor was recycled by the addition of *iso*-propanol.[4]

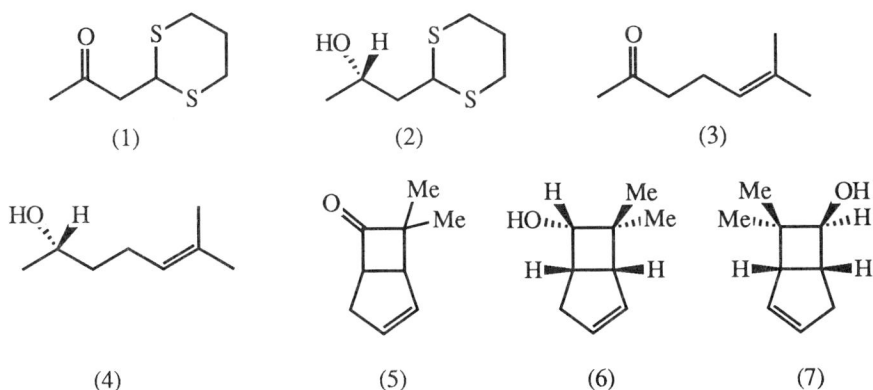

Ketones possessing pre-existing chiral centre(s) can be reduced to give diastereoisomeric products or, alternatively, the reduction can take place in enantioselective fashion. For example the bicyclic ketone (5) is reduced using the fungus *Mortierella rammaniana* to give the diastereomers (6) (20% yield; 80% e.e.) and (7) (20% yield; > 95% e.e.): the same racemic ketone is partially reduced by 3α,20β-hydroxysteroid alcohol dehydrogenase to give the alcohol (6) (30% yield; > 97% e.e.) and recovered optically active ketone (employed NADH as the co-factor and recycling the co-factor using horse liver alcohol dehydrogenase and ethanol). The optically active alcohols (6) and (7) and the optically active ketone (5) are useful synthons for the preparations of the chemo-attractant leukotrienes and the pheromone (+)-eldanolide.[5]

The enzyme-controlled reduction of β-diketones has been investigated; 2,2-disubstituted cyclopentane-1,3-diones have been popular substrates and bakers' yeast is often the micro-organism of choice for this bioconversion. For example Brooks *et al.*[6] have converted the 2,2-disubstituted cyclo-pentane-1,3-dione (8) into the hydroxyketone (9) *en route* to coriolin.

Similarly the reduction of β-ketoesters such as various alkyl 3-oxobutan-
oates has received attention.   The biotransformation of ethyl 3-oxobutan-
oate into ethyl 3-(S)-hydroxybutanoate using bakers' yeast is well
documented[7] and the stereoselectivity of the yeast reduction of various
4-substituted 3-oxobutanoates can be predicted with a reasonable degree of
accuracy.[8]   The reduction of 2-substituted 3-oxobutanoates takes place with
concommitant epimerization such that a high yield of one diastereoisomer of
the product can be obtained.   For instance the racemic ketoester (10) is
reduced to the hydroxy alkanoate (11) (92% e.e.).[9]

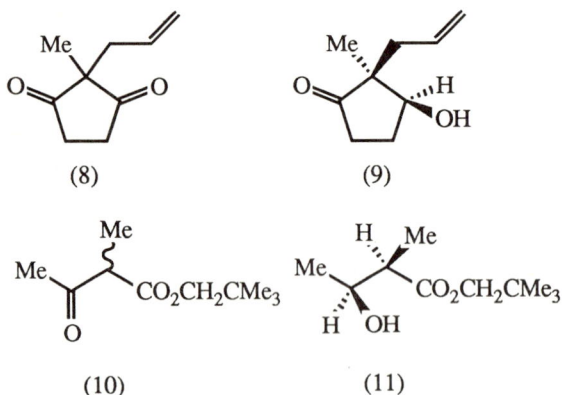

(8)   (9)

(10)   (11)

2.2   Reduction of αβ–Unsaturated Carbonyl Compounds

αβ–Unsaturated aldehydes, ketones and esters suffer reduction of the
carbon–carbon double bond using yeast and other microorganisms.   The
aldehyde (12) is reduced to the alcohol (13) using yeast (72% yield; > 99%
e.e.)[10] while the diketone (14) is transformed into the cyclohexanedione
(15) (55% yield;   90% e.e.) using *Aspergillus niger*.[11]   The conjugated
double bond is reduced in the presence of a non–conjugated alkene unit
within the same molecule as illustrated by the formation of the octenol
(17) (25% yield;   100% e.e.) from the substrate (16).[12]   αβ–Unsaturated
esters can be reduced using enoate reductase enzymes.   These enzymes, and
the associated relay system set up to use dioxygen as the oxidant, are not
easy to use in a non–specialist laboratory: nevertheless some very
impressive reactions have been reported, for example the reduction of the
allene carboxylic acid (18) to the αβ–unsaturated acid (19) in excellent
yield and in good enantiomeric excess.[13]

(12)

(13)

(14)

(15)

(16)

(17)

(18)

(19)

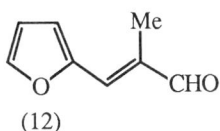

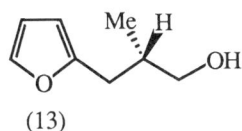

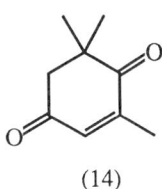

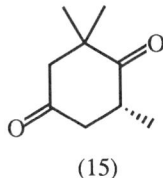

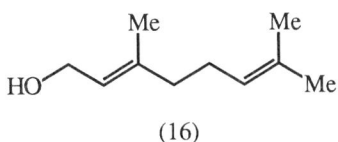

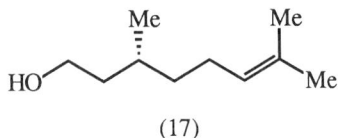

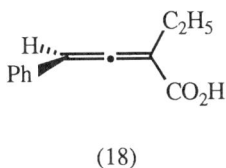

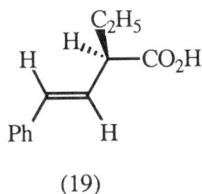

3.   ENZYME CATALYSED OXIDATION REACTIONS

3.1   Conversion of Alcohols into Aldehydes and Ketones

The transformation of either a primary alcohol into an aldehyde or a
secondary alcohol into a ketone using an enzyme as a catalyst, has not been
utilized in many laboratories.   This is because, *inter alia*, the recycling
of $NAD(P)^+$ is not as straightforward a process as that of recycling the
reduced co-factor.   However one excellent example of this enzyme-catalysed
process is described in the seminal work of J.B. Jones who used horse liver
alcohol dehydrogenase (with NAD^+ as co-factor) to oxidize the diol (20) to
the hydroxy aldehyde (21) (e.e. = 100%).[14]   The oxidation of one of the two
hydroxyl groups in compound (20) to give an optically active product
illustrates one of the prime advantages of using a chiral enzyme as a
catalyst rather than a more conventional non-chiral chemical oxidant.
Jones *et al.* found that the aldehyde (21) was further oxidized *in situ* to
give a key intermediate in the synthesis of grandisol.

3.2   Oxidation of Alicyclic Compounds

The regioselective hydroxylation of alicyclic compounds at a position
remote from pre-existing functional groups is a particularly interesting
bio-conversion.   The conversion of progesterone into 11-hydroxyprogesterone
and the importance of this series of compounds as anti-inflammatory agents
has been detailed elsewhere.[15]   The hydroxylation of the azaspiroundecane
(22) to produce the compound (23) in 81% yield using *Beauvaria*
sulphurescens[16] furnishes another example of this type of reaction.

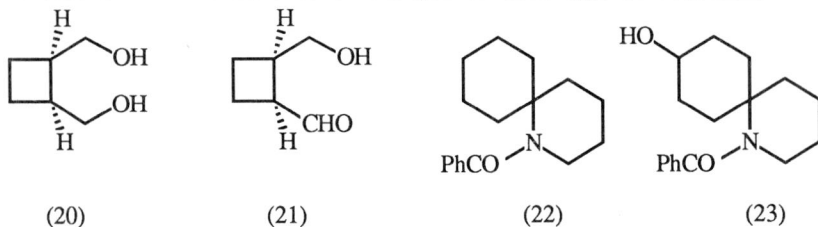

| (20) | (21) | (22) | (23) |

The conversion of a cyclic ketone into a lactone can be achieved by
using an isolated mono-oxygenase enzyme[17] or by using whole cells.   Thus
Acinetobacter NCIR 9871 converts the racemic dihaloketone (24) into the
optically active lactone (25) (36% yield;   > 95% e.e.) and recovered
optically active ketone (30% yield;   > 95% e.e.).   The latter compound was
converted into a carbocyclic nucleoside with activity against human
immunodeficiency virus (HIV).[18]

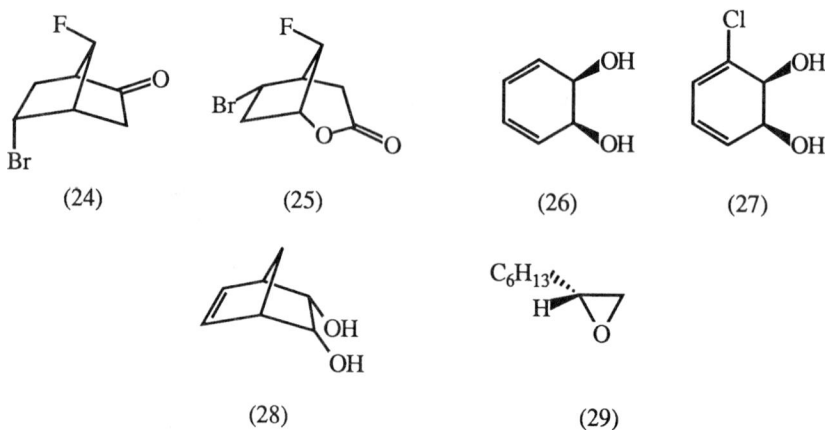

| (24) | (25) | (26) | (27) |

| (28) | (29) |

3.3 Oxidation of Aromatic Compounds

The regioselective oxidation of aromatic compounds to produce selected phenols has been the subject of some research.[19] However, the emphasis on the area of the oxidation of benzene derivatives has shifted recently, with a considerable amount of interest now being focussed on the formation of cyclohexadiene-1,2-diol and derivatives. The parent compound (26) is available from benzene using *Pseudomonas* spp[20] and this compound has been converted into natural products such as an inositol triphosphate.[21] Substituted benzenes give optically active synthons; thus chlorobenzene provides the diol (27) and this has been converted into protected (D)- and (L)-erythrose as an illustration of the potential usage of this class of compounds.[22] *Pseudomonads* are also available which are able to catalyse the oxidation of benzoic acid derivatives.[23] Some organisms which have been used to oxidize benzene to cyclohexadienediol can also catalyse the oxidation of norbornadiene to the diol (28).[24]

3.4 Miscellaneous Enzyme-Catalysed Oxidation Reactions

The stereocontrolled oxidation of an alkene to form an optically active epoxide would be a valuable biotransformation. Only a few examples are mentioned in the literature: the conversion of octene into the epoxide (29) is one of the examples.[25] Sulphoxides are prepared from sulphides using, as catalysts, enzymes that are present in whole-cell systems. Excellent enantiomeric excesses of products have been reported but over-oxidation to the corresponding sulphone can present a problem.[26]

4. CONCLUSIONS

Only a handful of enzyme-catalysed reduction reactions and a similar number of enzymic oxidation reactions have been scaled up to provide bio-transformation products on a large scale. There can be no doubt that enzyme-catalysed oxidation and reduction reactions are more difficult to perform than hydrolysis reactions principally because of the problems associated with co-factor recycling. More research is needed to provide more simple systems that are reliable and can be used by non-specialists. This research is currently underway and significant progress can be expected over the next decade.

REFERENCES

1. H.G. Davies, R.H. Green, D.R. Kelly, and S.M. Roberts, "Biotransformations in Preparative Organic Chemistry: the Use of Isolated
   Enzymes and Whole Cell Systems in Synthesis", Academic Press, London,
   1989.

2. S. Butt and S.M. Roberts, Chem. in Brit., 1987.

3. P.-F. Deschenaux, T. Kallimopoulos, and A. Jacot-Guillarmod, Helv.
   Chim. Acta, 1989, 72, 1259.

4. E. Keinan, E.K. Hafeli, K.M. Seth, and R. Lamed, J. Am. Chem. Soc.,
   1986, 108, 162.

5. H.G. Davies, S.M. Roberts, B.J. Wakefield, and J.A. Winders, J. Chem.
   Soc., Chem. Commun., 1985, 1166.

6. D.W. Brooks, H. Mazdiyarni, and P. Sallay, J. Org. Chem., 1985, 50,
   3411.

7. Y. Naoshima, A. Nakamura, T. Nishiyama, T. Haramaki, M. Mende, and
   Y. Munakata, Chem. Lett., 1989, 1023; see also Y. Naoshima and
   Y. Akakabe, J. Org. Chem., 1989, 54, 4237 for the use of immobilised
   Nicotiana tabacum for this biotransformation.

8. C.J. Sih and C.S. Chen, Angew. Chem. Int. Ed. Engl., 1984, 23, 570.

9. K. Nakamura, T. Miyai, A. Nagar, S. Oka, and A. Ohno, Bull. Chem. Soc.
   Japan, 1989, 62, 1179.

10. C. Fuganti, P. Grasselli, S. Servi, and H.E. Högberg, J. Chem. Soc.,
    Perkin Trans. 1, 1988, 3061.

11. Y. Yamazaki, Y. Hayashi, N. Hori, and Y. Mikami, Agric. Biol. Chem.,
    1988, 52, 2919.

12. P. Gramatica, P. Manitto, D. Monti, and G. Speranza, Tetrahedron,
    1986, 42, 6687; presumably 3,7-dimethylocta-2,6-dienal is transiently
    formed in this biotransformation.

13. H. Simon, H. Gunther, J. Bader, and W. Tischer, Angew. Chem. Int. Ed.
    Engl., 1981, 20, 861.

14. J.B. Jones, M.A.W. Finch, and I.J. Jakovac, Can. J. Chem., 1982, 60,
    2007.

15. "Medicinal Chemistry: The Rôle of Organic Chemistry in Drug Research"
    eds. B.J. Price and S.M. Roberts, Academic Press, Orlando, 1985.

16. W. Carruthers, J.D. Prail, S.M. Roberts, and A.J. Willetts,
    unpublished results.

17. C.T. Walsh and Y.-C.J. Chen, Angew. Chem. Int. Ed. Engl., 1988, 27,
    333.

18. M.S. Levitt, R.F. Newton, S.M. Roberts, and A.J. Willetts, J. Chem. Soc., Chem. Commun., submitted for publication.

19. see for example, J.C. Cox and J.H. Golbeck, Biotechnol. Bioeng., 1985, 27, 1395.

20. D.G.H. Ballard, A. Courtis, I.M. Shirley, and S.C. Taylor, J. Chem. Soc., Chem. Commun., 1983, 954.

21. S.V. Ley, M. Parra, A.J. Redgrave, F. Sternfeld, and A. Vidal, Tetrahedron Lett., 1989, 30, 3557.

22. T. Hudlicky, H. Luna, J.D. Price, and F. Rulin, Tetrahedron Lett., 1989, 30, 4053.

23. S.J.C. Taylor, D.W. Ribbons, A.M.Z. Slawin, D.A. Widdowson, and D.J. Williams, Tetrahedron Letters, 1987, 28, 6391.

24. P.J. Geary, R.J. Pryce, S.M. Roberts, G. Ryback, and J.A. Winders, J. Chem. Soc., Chem. Commun., in press.

25. M.J. de Smet, B. Witholt, and H. Wynberg, J. Org. Chem., 1981, 46, 3128.

26. H. Ohta, Y. Okamoto, and G.-I. Tsuchihashi, Agric. Biol. Chem., 1985, 49, 671 and 2229.

CHEMO-ENZYMATIC PRODUCTION METHODS FOR OPTICALLY PURE INTERMEDIATES OF POTENTIAL COMMERCIAL INTEREST

Q.B. BROXTERMAN*, W.H.J. BOESTEN, J. KAMPHUIS, M. KLOOSTERMAN,
E.M. MEIJER AND H.E. SCHOEMAKER
DSM Research, Bio-organic Chemistry Section
P.O. Box 18, 6160 MD Geleen, The Netherlands

ABSTRACT

Possible opportunities for bio-catalytic production methods for enantiomerically pure synthetic intermediates depend on the chemo-economical environment of the enzymatic step. This is elucidated on the basis of a number of criteria that link the bio-catalytic step to its chemical environment. The concept of the chemo-enzymatic synthesis tree is presented as an example of a mixed short- and long-term strategy for identifying opportunities for biotransformations. This concept rests on two pillars. Firstly, a set of related primary products can be obtained from a set of related starting materials and a single enzyme (relaxed substrate specificity). Secondly, a broad range of secondary products are obtained from the primary products by classical (non-enzymatic) chemistry. The concept is illustrated by the L-specific aminopeptidase-catalyzed preparation of L-and D-α-hydrogen amino acids. A number of unnatural amino acids are put on stage as examples of primary products. Enantiomerically pure α-amino alcohols are chosen to represent the secondary products. Finally, a related chemo-enzymatic process for the preparation of α,α-dialkylamino acids is also presented.

INTRODUCTION

Appreciation of opportunities in biotransformations strongly depends on the position of the observer in the 'bio-field'. When narrowing down the large area of biotransformations to chemo-enzymatic production methods for synthetic intermediates, the appreciation depends on the fulfilment of a set of criteria. Some of the more important criteria are:

1. The bio-catalytic step (including work-up) should be a cost-effective
   and technically feasible production method yielding material of the
   required specifications.
2. The position of the bio-catalytic step in the synthetic route planned
   for the target intermediate should be logical.
3. The substrate for the bio-catalytic step preferably corresponds with
   existing production, or else can be obtained cheaply and easily.
4. Analysis of the relationship between own and world capacity and processes and the world market (alternative intermediates for same
   endproducts to be included) should show a positive result.

Phrased somewhat differently the appreciation would depend on the
use of an economically attractive starting material in a, logically
planned, technically sound bio-catalytic step for an intermediate with a
clear-cut market.
Practical application of these criteria means that in a very real sense
the chemo-economical environment of the bio-catalytic step is at least as
important in determining the opportunity as the bio-catalytic step
itself. Examples will be given later on.

An industrial starting point for determining whether of not a perceived biocatalytic step presents an opportunity is the establishment of
a commercially relevant target molecule (e.g. an enantiomerically pure
synthetic intermediate). Establishment of the target molecule(s) before
doing extensive R & D is an example of relatively short-term strategy. It
means that serious work is only started provided the path from start to
finish can be surveyed. More often than not target molecules from this
category are suddenly 'in the air'. Consequently, all of a sudden one
encounters heavy competition and intense activity in these area.
Examples in this category are:
- enantiomerically pure α-chloropropionic acid and analogues (for optically active phenoxy herbicides),
- enantiomerically pure epichlorohydrin of equivalent glycerol-like
  C_3-synthons (for enantiomerically pure β-blockers).
Many fine chemo-enzymatic processes have been developed in this category,
but a good number of opportunities remain uncovered because the size of
the pond in which to fish is limited.

Interesting market niches can be disclosed by combining the outcome of
the bio-catalytic step with additional non-enzymatic chemistry.
The result of this combination leads to a tremendous expansion of the
synthetic potential of the bio-catalytic step (viz. the chance of identifying an opportunity). This approach is an example of long-term strategy because now, together with the immediate targets, additional technology is developed for which commercial interesting targets are defined
on the way.

THE CHEMO-ENZYMATIC SYNTHESIS TREE; PRIMARY AND SECONDARY PRODUCTS

A particulary strong chemo-enzymatic synthesis tree will grow if the
biocatalyst has a relaxed substrate specificity; this means that a number
of related starting materials all undergo the bio-catalytic step with
comparable efficiency and (stereo)selectivity.
The outcome of the bio-catalytic step can be seen as the set of primary
products. A multitude of secondary products result from the reactions of
primary products with a set of classical (non-enzymatic) reagents.
The whole process from starting material to secondary product is repre-
sented in Figure 1.

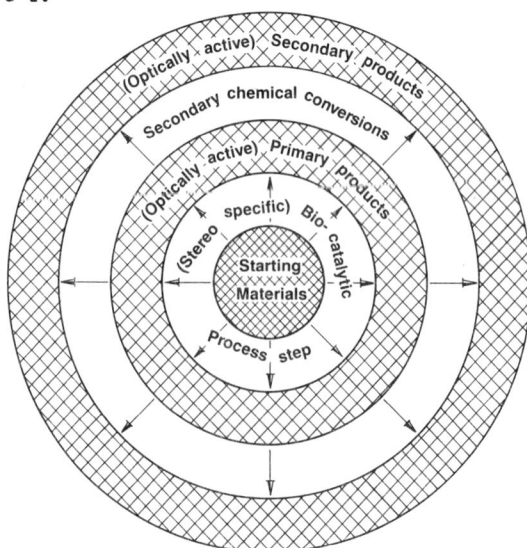

Figure 1. The chemo-enzymatic synthesis tree.

An example of the latter approach outlined above is the DSM chemo-
enzymatic production method of enantiomerically pure amino acids and
derivates. The key-step in this chemo-enzymatic process is the stereospecific hydrolysis of only the L-amino acid amide to the L-amino acid
(Scheme 1) [1 - 3]. The high stereoselectivity and the ease of operation
are important advantageous aspects of this bio-catalytic step. Generally,
both L-acid and D-amide are isolated with an ee above 98 %.
Permeabilized whole cells of *Pseudomonas putida* ATCC 12633, containing
the L-aminopeptidase are used as such. However, some non bio-catalytic
features of the process are also of importance.
On one hand, separation of L-acid from D-amide is very much facilitated
by formation of the insoluble Schiff base of the D-amide, while on the
other hand racemization and recycling possibilities of either L- or D-
amide are built into the process scheme (Scheme 1). Table 1 shows that
the L-aminopeptidase from *Pseudomonas putida* indeed shows relaxed
substrate specifity.

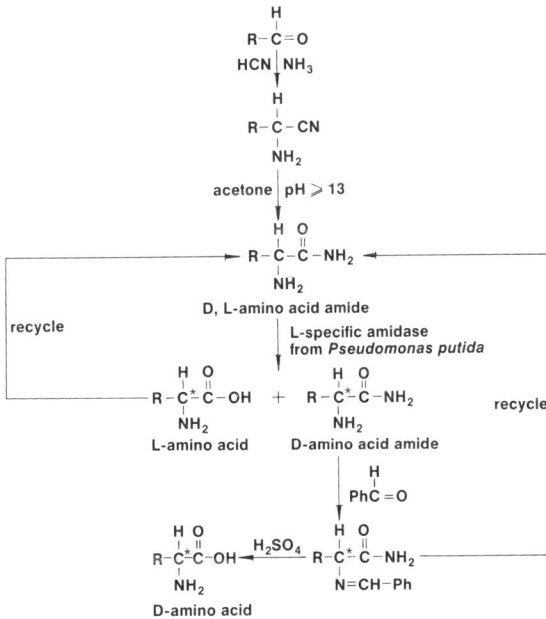

Scheme 1. Chemo-enzymatic synthesis of enantiomerically pure amino acids
by aminopeptidase technology

TABLE 1

The relaxed substrate specificity of *Pseudomonas putida*.

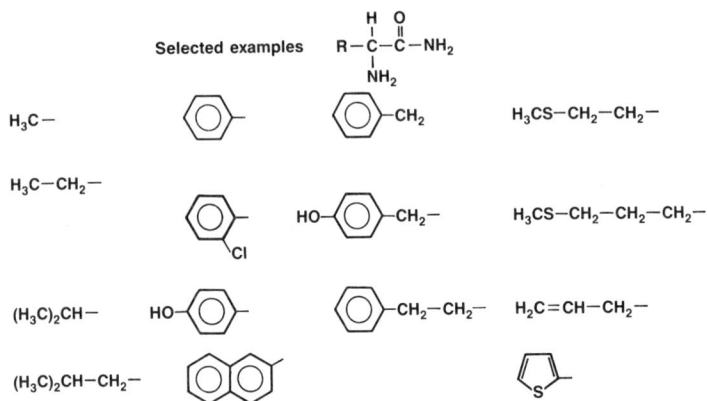

Clearly this process offers opportunities for the chemo-enzymatic production of both L- and D-amino acids.

PRIMARY PRODUCTS; EXAMPLE: UNNATURAL AMINO ACIDS

Apart from the evident commercially attractive amino acids, new targets
can be identified and thus new opportunities created for this bio-
catalytic process by developing appropiate derivatization chemistry for
the enantiomerically pure amino acids.

The potential use of homophenylalanine as a chiral building block in
the synthesis of a number of ACE inhibitors (Figure 2) constitutes a fine
example.

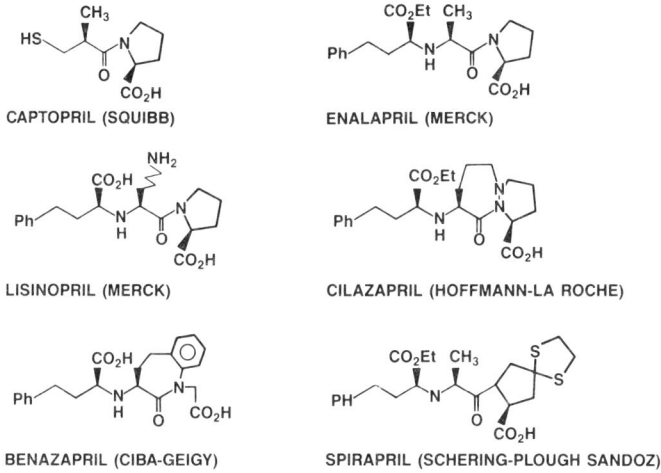

Figure 2. Structures of various ACE inhibitors.

Homophenylalanine (HPhe), being an unnatural amino acid, cannot be obtained by fermentation. However, this amino acid has been resolved on a commercial scale using the DSM chemo-enzymatic approach. The substrate for the enzymatic step is easily prepared from *i.a.* benzaldehyde, hydrogen cyanide and ammonia (remember criterium 3): see Scheme 2. Both L- and D-homophenylalanine were obtained by this route; this is of interest since they are both commercially interesting starting materials.

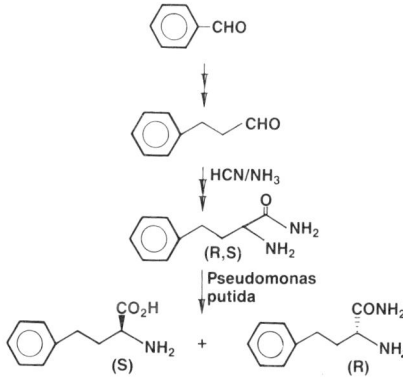

Scheme 2. Chemo-enzymatic preparation of L- and D-homophenylalanine.

This can be illustrated with the example of Enalapril, a product of
Merck, Sharp & Dohme, which is an important representative of the class
of ACE inhibitors. MSD-scientists published a production method consisting of the reductive amination of ethyl 4-phenyl-2-keto-butyrate with
L-Ala-L-Pro [4, 5]. Crude Enalapril with an ee of 74 % results from this
process [4]. This method entails an additional crystallization of its
maleic acid salt.

Some alternative commercially interesting routes have been developed at
Andeno/DSM [6, 7]. Coupling of ethyl L-homophenylalanate with a derivative of D-lactic acid (D = R) affords, after deprotection, the key intermediate 1 (Scheme 3). Compound 1 gives Enalapril on reaction with L-
proline, but it can also be used for structurally related ACE inhibitors
(e.g. Spirapril, Figure 2).

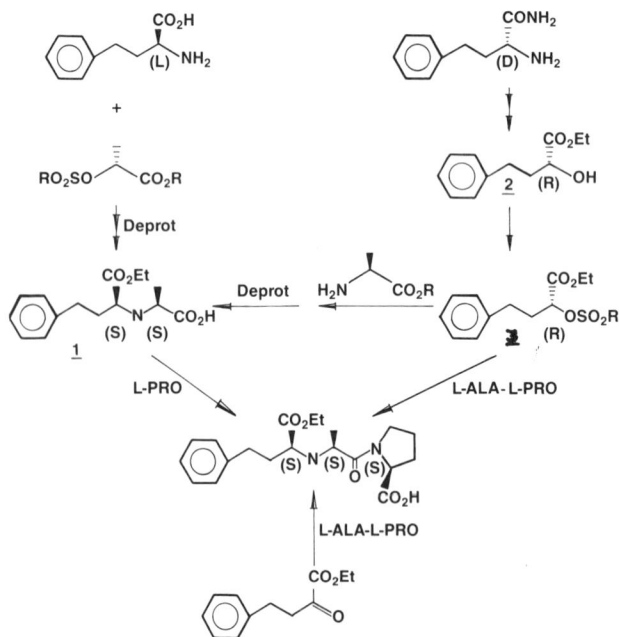

Scheme 3. Synthesis of Enalapril from L- or D-homophenylalanine.

On the other hand, D-homophenylalanine amide could be converted effi-
ciently in a few steps to the corresponding optically pure α-hydroxy
ester 2. Intermediate 2 is transformed into 1 in three steps, using an
L-alanine ester as the second chiral building block.

A third and very short route, starting from D-homophenylalanine involves reaction of intermediate $\underline{3}$ with L-Ala-L-Pro. An interesting aspect of this sequence is that other ACE inhibitors which contain homophenylalanine, but not next to L-alanine, can also be prepared from $\underline{3}$ (e.g. Cilazapril, Benazapril, Lisinopril; see Figure 2).

Whether or not these alternative routes represent opportunities for the aminopeptidase process depends in part on the chemo-economical environment. In terms of the major raw materials this is illustrated in Scheme 4. The reagents are shown, in this case, with their biotechnological production methods.

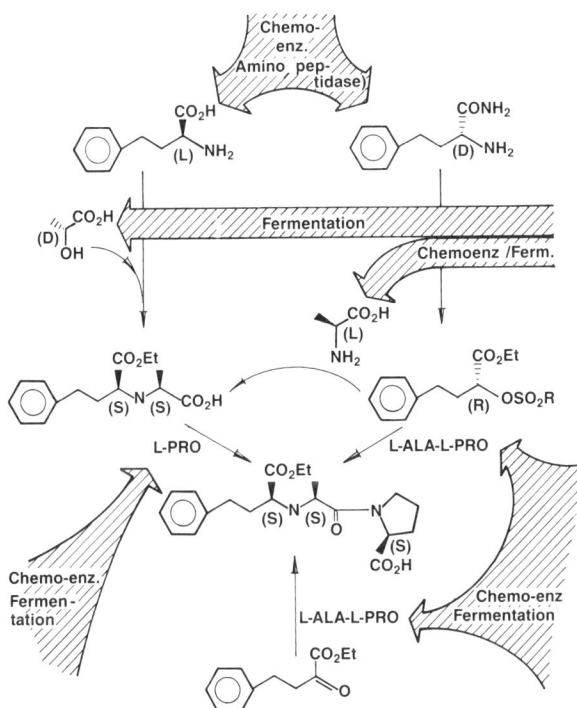

Scheme 4. Opportunities for biotransformations; an example of relevant interplay

The commercial viability of the chemo-enzymatic production method in the top of Scheme 4, as far as this application is concerned, depends on the price of the (biotechnologically) produced materials further down.

Note that in this case, like in many others, there is a complex interplay
which influences opportunities for various biotechnologically producable
compounds. Table 2 summarizes the materials to be compared.

TABLE 2
Major raw materials for Enalapril production

L-HPhe-1-E	D-HPhe-2-3-1-E	D-HPhe-2-3-E	Reductive amination-E
L-HPhe D-lactic acid L-Pro	D-HPhe L-Ala (2 → 3 → 1) L-Pro	D-HPhe L-Ala-L-Pro (2 → 3)	Ethyl 4-phenyl-2-ketobutyrate L-Ala-L-Pro

HPhe = Homophenylalanine; E = Enalapril

The resolution step in all routes starting from L- or D-homophenylalanine is early in the overall scheme. This avoids wasting reagents etc.
on the wrong stereoisomer which is, as such,an application of
criterium 2.

Trifunctional amino acids constitute a class of unnatural amino acids
which has currently our special interest. On account of their third func-
tionally a multitude of enantiomerically pure amino acid derivatives can
be prepared from this particular class of synthetically versatile chiral
building blocks. D-cycloserine (see figure 4) is a commercial example: it
is after all the internal amide of the D-isomer of O-aminoserine.
Due to the relaxed substrate specificity of the L-aminopeptidase of
Pseudomonas putida it was possible to prepare enantiomerically pure trifunctional amino acids by this method (Scheme 1). Figure 3 shows examples
of amino acids with an additional unsaturation (allylglycine), a sulphur
moiety (homomethione) [8], or an oxygen moiety (O-benzyl bishomoserine)
which were recently obtained optically pure by aminopeptidase
technology.

Optically active trifunctional amino acids

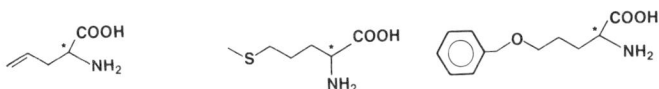

Figure 3. Enantiomerically pure trifunctional amino acids obtained by
aminopeptidase technology

SECONDARY PRODUCTS; EXAMPLE: α-AMINO ALCOHOLS

The possibilities to broaden the scope of the aminopeptidase technology
by further derivatization of enantiomerically pure amino acids are
enormous. A few chosen examples are shown in Scheme 5. As can be seen
from Scheme 5 optically pure L- and D-α-amino alcohols, α-halocarboxylic
acids, α-hydroxy carboxylic acids and N-hydroxy-α-amino acid amides have
been obtained from the corresponding L- or D-amino acids (or amides) [9].

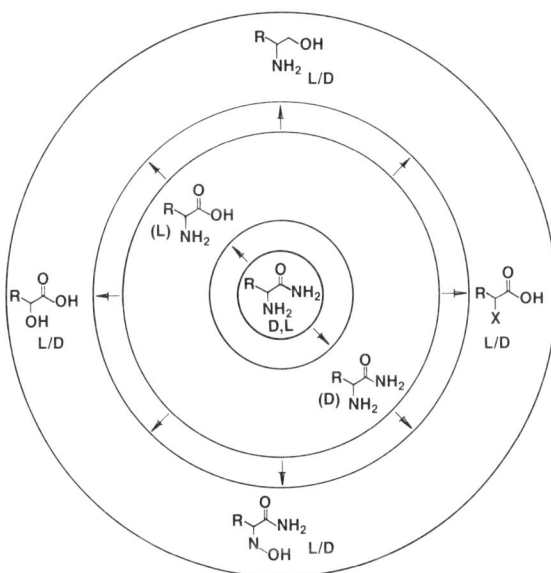

Scheme 5. Some secondary products from the L-aminopeptidase based chemo-
enzymatic synthesis tree.

This general 'catalogue' of opportunities will be highlighted with a few examples from the class of enantiomerically pure α-amino alcohols.
To begin with a comparison will be made between the L-aminopeptidase technology and a lipase-based technology. Two of the numerous methods to prepare optically enriched (or pure) α-amino alcohols are the chemical reduction of optically pure amino acids and the lipase catalyzed saponification of esters of amino alcohols.

The lipase catalyzed stereospecific hydrolysis of dibutanoyl phenylglycinol and dibutanoyl phenylalaninol is shown in Scheme 6 [10].

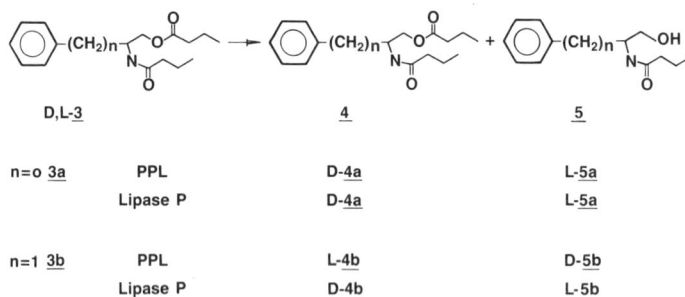

D,L-3		4	5
n=o 3a	PPL	D-4a	L-5a
	Lipase P	D-4a	L-5a
n=1 3b	PPL	L-4b	D-5b
	Lipase P	D-4b	L-5b

Scheme 6. Lipase catalyzed hydrolysis of dibutanoyl derivates of racemic amino alcohols.

Treatment of dibutanoyl-D,L-phenylglycinol (3a) with a lipase preparation from porcine pancreas resulted in the formation of the corresponding D-ester 4a and L-alcohol 5a. A reasonable enantiomeric excess was observed: at 49 % conversion 88 % c.q. 91 % ee. With lipase from a *Pseudomonas* strain the stereoselectivity was slightly less, but a higher initial velocity V_0 was measured (i.e. 0,27 versus 0,02 U/mg solid).
Upon treatment of dibutanoyl-D,L-phenylalaninol 3b with a lipase preparation from a *Pseudomonas* strain again the formation of D-ester (D-4b) and L-alcohol (L-5b) was observed, albeit with lower enantioselectivity E (E = 10). In contrast, PPL-catalyzed hydrolysis of 3b yielded preferentially L-ester (L-4b) and D-alcohol (D-5b). The initial velocities V_0 for the hydrolysis of 3b by lipases from *Pseudomonas* and porcine pancreas were determined and found te be 0,43 c.q. 0,30 U/mg solid.
Evaluation of these experiments leads to the conclusion that reduction of optically pure amino acids obtained by aminopeptidase technology is superior to lipase catalyzed ester hydrolysis.

This conclusion is based on the higher ee's (above 98 % versus below 91 %), the more cost-effective reduction of only enantiomerically pure instead of racemic amino acid (a method to prepare it), and the possibility of recycling unwanted amino acid amide. The first of these 3 remarks boils down to a validation of criterium 1, and the second a validation of criterium 3.

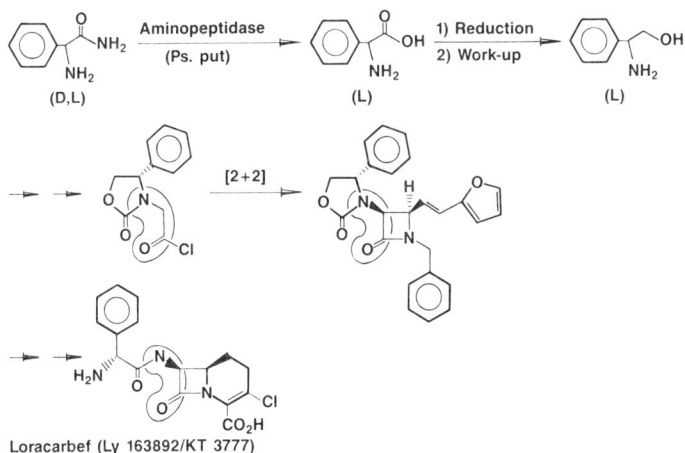

Scheme 7. A possible opportunity for L-phenylglycinol.

A possible application of L-phenylglycinol is its use as efficient chiral auxiliary in an asymmetric [2 + 2] addition reaction eventually leading to Loracarbef. This compound is an experimental antibiotic developed by Lilly and prepared on kg-scale in order to obtain sufficient material for clinical trials (Scheme 7) [11].

This scheme illustrates that a possible opportunity for chemo-enzymatic production of L-phenylglycine depends, as far as this application is concerned, on a development chemically quite remote from the bio-catalytic step: the efficiency of chiral induction of the L-phenylglycinol derivative in the [2 + 2] cycloaddition reaction. Note that L-phenylglycine excellently fulfils criterium 3 (being based on benzaldehyde, hydrogen cyanide and ammonia).

The use of enantiomerically pure α-amino alcohols is not restricted to pharmaceutical chemistry.

The example of 4-substituted 1,3,2-oxazaphospholidine-2-sulfides stems from the agrochemical area (Scheme 8). These compounds are derived from α-amino alcohols.

It was discovered that the members of this class of compounds prepared from L-valinol (R = iPr) or L-leucinol (R = iBu) are more active insecticides than either their enantiomers or analogues with a different substituent at position 4 [12 - 14].

L-Valinol / L-Leucinol

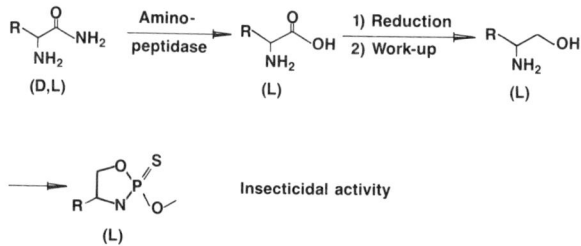

R= iPr, iBu

Scheme 8. A possible opportunity for L-valinol or L-leucinol.

Enantiomerically pure L-2-amino-1-butanol is another interesting amino alcohol. It can be used either as a resolving agent or as a chiral building block for Ethambutol, a tuberculostatic produced by Lederle [15, 16].

A number of the known strategies to produce L-2-amino-1-butanol are collected in Scheme 9. Racemic amino alcohol has been prepared from nitropropane in two steps and from 2-amino-butyric acid or its ester. Classical resolution of racemic amino alcohol (e.g. with tartaric acid) is an option [15, 16].

Francalanci et. al. reported two different lipase-based transformations leading to products with an enantiomeric excess for the (S)-isomer above 95 % [17]: enzymatic hydrolysis and enzymatic transesterification (Scheme 9). The reduction of enantiomerically pure 2-amino butyric acid, obtained by e.g. L-aminopeptidase technology is yet another way to S-amino alcohol.

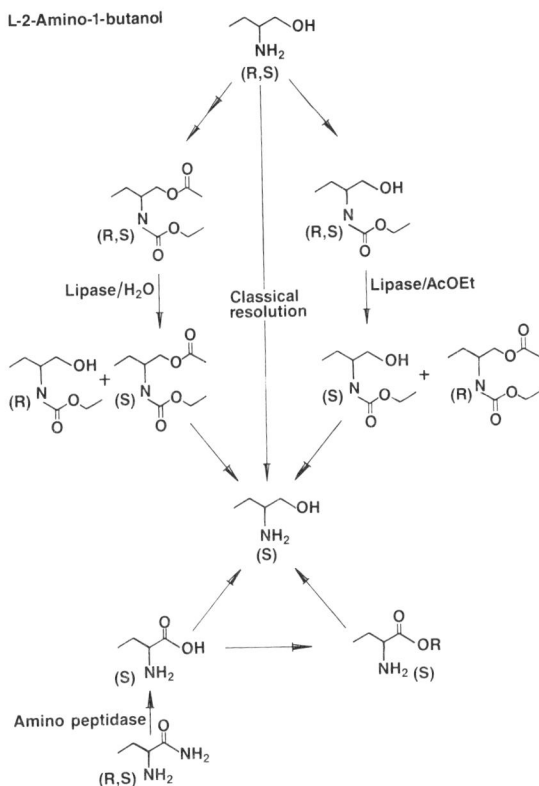

Scheme 9. Some different synthetic routes towards L-2-amino-1-butanol.

In order to determine where the opportunity lies a comparison between chemical resolution, the lipase-based strategy and the aminopeptidase technology is mandatory (Scheme 10).

In this case it seems that the ee in the lipase route does not present a problem. However, the considerable amount of activating/deprotecting chemistry centred round the bio-catalytic step is detrimental. Although simple reagents are sufficient, it takes three steps from racemic to S-amino alcohol. This is evidently a violation of criterium 2.

Combined with the relatively difficult racemisation of amino alcohols and the lack of specific advantages it renders the lipase route less active.

The other two routes have both advantages and are economically much
closer to each other. The classical resolution uses cheap racemic amino
alcohol (from nitropropane) but probably suffers from complex recycle
streams and difficult racemisation of R-amino alcohol.

The aminopeptidase route involves a relatively expensive reduction (but
only on enantiomerically pure acid). However, production of S-amino acid
is efficient with possibilities for facile racemisation of D-amino acid
amide.

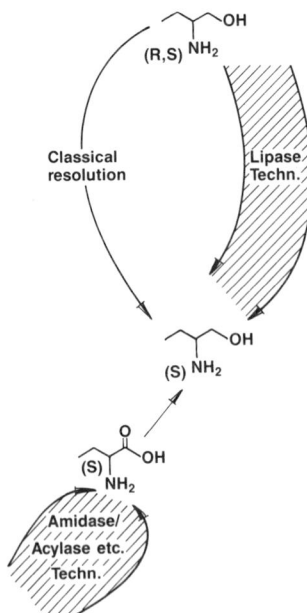

Scheme 10. Relationships between different strategies to produce
L-2-amino-1-butanol

OPPORTUNITIES FOR CHEMO-ENZYMATIC PRODUCTION METHODS; GENERAL REMARKS

As stated and illustrated before the chemo-economical environment of the
bio-catalytic step is sometimes of decisive importance. It is therefore
difficult, for example in the area of enantiomerically pure amino acids,
to make general statements about the relative economics of

chemoenzymatic, fermentation, and classical resolution procedures.
The competition in natural L-amino acid production tends to be between
chemo-enzymatic and fermentation processes. In contrast, competition for
the 'unnatural' D-isomers of natural amino acids and for 'completely
unnatural' amino acids (e.g. L- and D-homophenylalanine) tends to be between chemo-enzymatic and classical resolution protocols. The limitation
of these general statements however, is nicely illustrated with the
6 D-amino acids collected in Figure 4.

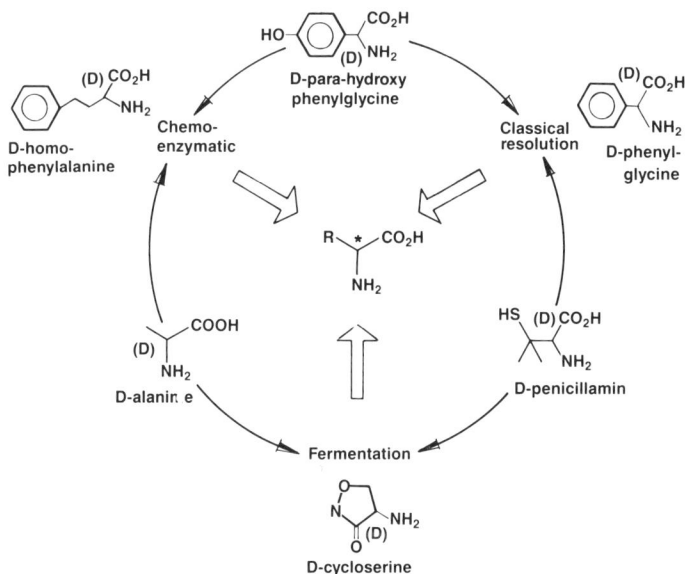

Figure 4. Production methods for some selected D-amino acids

For some of these D-amino acids only one serious production process is
known (D-homophenylalanine, D-phenylglycine [7], D-cycloserine [18]).
For other D-amino acids two different technologies are in use
(D-parahydroxyphenylglycine [19]), or have been used (D-pencillamin [20])
on a commercial scale. For D-alanine two different technologies are under
development [21].
Table 3 summarizes the (potential) use of these D-amino acids as chiral
building block or as chiral endproduct itself.

TABLE 3
(Potential) use of various D-amino acids.

Amino acid	Function	(potential) use in
D-phenylglycine	intermediate	antibiotic
D-parahydroxyphenylglycine	intermediate	antibiotic
D-homophenylalanine	intermediate	ACE inhibitors
D-alanine	intermediate	artificial sweetener
D-cycloserine	endproduct	tuberculostatic
D-penicillamin	endproduct	agent against rheumatoid arthritis

It is clear that both chemo-enzymatic production methods as well as classical resolution suffer from the requirement of recycle streams which are dictated by the necessity to racemise the unwanted isomers. After all, only in an ideal chemo-industrial world would there be an outlet for equal amounts of D- and L-isomers.

For the chemo-enzymatic route it means that in-situ racemisation during the bio-catalytic step represents an enormous advantage. A well-known example is the racemisation of the L-isomer of the hydantoin of para-hydroxyphenylglycine during the action of D-hydantoinase on the D-isomer [19]. Opportunities for a general chemo-enzymatic process like the amino-peptidase technology would therefore arise from a combination of L-aminopeptidase/racemase and/or D-aminopeptidase/racemase (Scheme 11, optimisation of criterium 1).

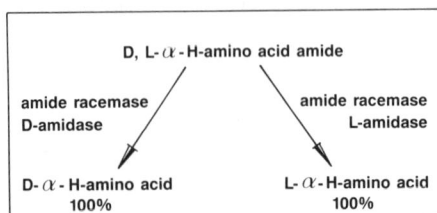

Scheme 11.

In-vivo protein engineering is a way to tackle this problem. It means selection of micro-organisms by goal-oriented growth conditions in continuous cultivation. Recently we have been able to obtain a mutant of the original *Pseudomonas putida* strain (with L-aminopeptidase) which contains additional D-aminopeptidase and racemase activity [22]. Selective blocking of either the L- or the D-aminopeptidase activity then yields the desired bio-catalyst.

CHEMO-ENZYMATIC PRODUCTION OF α,α-DISUBSTITUTED AMINO ACIDS

The synthesis of enatiomerically pure α,α-disubstituted amino acids by chemo-enzymatic methods is an example of a new development in the area of optically pure intermediates. These compounds are mainly of interest as (building blocks in) pharmaceuticals, L-α-methyl DOPA being the prime commercial example. However, new applications will arise from the usage of enantiomerically pure α,α-disubstituted amino acids as building blocks in the synthesis of analogues of bio-active peptides. Their use is of relevance because incorporation yields peptide analogues in which specific conformations tend to be frozen and enzymatic degradation processes tend to be slowed down dramatically.

The presence of an α-hydrogen in the amino acid amide proved to be essential for the enzyme activity of the aminopeptidase of *Pseudomonas putida*. Therefore, at DSM a new bio-catalyst from *Mycobacterium neoaurum* ATCC 25795 has been developed capable of stereoselective hydrolysis of α,α-disubstituted amino acid amides (Scheme 12) [23]. Note that also in this case the substrate for the enzymatic step is a precursor for the amino acid (criterium 2). Again a relaxed substrate specificity is observed (Table 4), allowing for a flexible response to emerging opportunities in the market.

$$R^1-\underset{\underset{O}{\|}}{C}-R^2 \xrightarrow[NH_3]{HCN} R^1-\underset{\underset{NH_2}{|}}{\overset{\overset{R^2}{|}}{C}}-CN \longrightarrow R^1-\underset{\underset{NH_2}{|}}{\overset{\overset{R^2}{|}}{C}}-\underset{\overset{\|}{O}}{C}-NH_2$$

D, L-α-alkyl substituted amino acid amide

L-specific amidase from
Mycobacterium neoaurum

recycle ——— $R^1-\underset{\underset{NH_2}{|}}{\overset{\overset{R^2}{|}}{C}}-\underset{\overset{\|}{O}}{C}-OH$ + $R^1-\underset{\underset{NH_2}{|}}{\overset{\overset{R^2}{|}}{C}}-\underset{\overset{\|}{O}}{C}-NH_2$

L-α-alkyl substituted
amino acid

D-α-alkyl substituted
amino acid amide

recycle ——— $R^1-\underset{\underset{NH_2}{|}}{\overset{\overset{R^2}{|}}{C}}-\underset{\overset{\|}{O}}{C}-OH$

D-α-alkyl substituted
amino acid

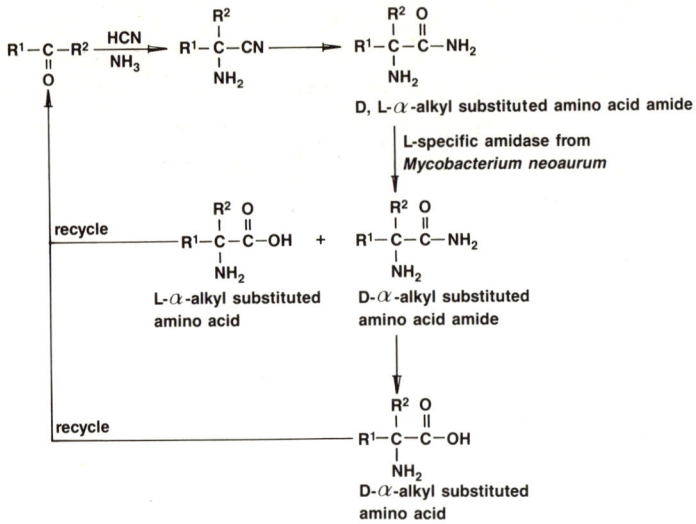

Scheme 12. Chemo-enzymatic production of enantiomerically pure
α,α-disubstituted amino acids.

TABLE IV

Substrate range used so far (and with success) in stereospecific
hydrolysis.

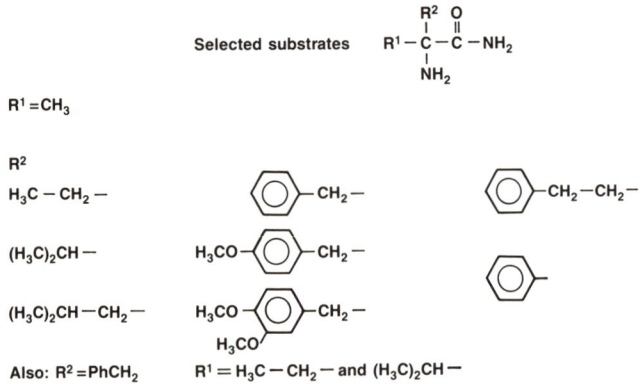

Selected substrates $\quad R^1-\underset{\underset{NH_2}{|}}{\overset{\overset{R^2}{|}}{C}}-\underset{\overset{\|}{O}}{C}-NH_2$

$R^1 = CH_3$

R^2

H_3C-CH_2- 　　　　　◯$-CH_2-$ 　　　　　◯$-CH_2-CH_2-$

$(H_3C)_2CH-$ 　　　H_3CO-◯$-CH_2-$

$(H_3C)_2CH-CH_2-$ 　H_3CO-◯$-CH_2-$ 　　　　　◯$-$
　　　　　　　　　　　H_3CO

Also: $R^2 = PhCH_2$ 　$R^1 = H_3C-CH_2-$ and $(H_3C)_2CH-$

CONCLUSIONS

The foregoing illustrates a strategy to identify opportunities for
biotransformations. It implies that in order to identify a real opportunity for bio-catalytic conversion it is mandatory to match a specific
catalytic step with market needs by creating a chemo-enzymatic synthesis
tree. Interestingly, although opportunities for amino acids and derivates
are more obvious and larger in number, the ICI biotechnology for producing a range of cis-1,2-dihydrocatechols is based on a similar philo-
sophy [24]. Coinciding with this strategy emphasis has been put on the
importance of a close watch of the chemo-economical environment of the
bio-catalytic step.

ACKNOWLEDGEMENT

We would like to acknowledge the important contributions from a large
number of colleagues both from DSM Research and Andeno.

REFERENCES

1. Meijer, E.M., Boesten, W.H.J., Schoemaker, H.E. and van Balken,
   J.A.M., Use of biocatalysts in the industrial production of speciality
   chemicals. In Biocatalysis in Organic Syntheses, eds. J. Tramper,
   H.C. van der Plas and P. Linko, Elsevier, Amsterdam, 1985,
   pp. 135-156.

2. Boesten, W.H.J., Dassen, B.H.N., Kerkhoffs, P.L., Roberts, M.J.A.,
   Cals, M.J.H., Peters, P.J.H., van Balken, J.A.M., Meijer, E.M. and
   Schoemaker, H.E., Efficient enzymatic production of enantiomerically
   pure amino acids. In Enzymes as Catalysts in Organic Synthesis. ed.
   M.P. Schneider, NATO ASI Series C, vol.178, Reidel Publishing Company,
   Dordrecht, 1986, pp. 355-360.

3. Boesten, W.H.J., US Pat. 3971700 (1976), British Pat. 1548032 (1976),
   US Pat. 4172846 (1979), US Pat. 4172846 (1979).

4. Blacklock, T.J., Shuman, R.F., Butcher, J.W., Shearin, Jr.,W.E.,
   Budavari, J. and Grenda, V.J., Synthesis of semisynthetic dipeptides
   using N-carboxyanhydrides and chiral induction on Raney Nickel.
   A method practical for large scale, J. Org. Chem., 53, 836-844 (1988).

5. Wyvratt, M.J., Tristam, E.W., Ikeler, T.J., Lohr, N.S., Joshua, H., Springer, J.P., Arison, B.H. and Patchett, A.A., Reductive amination of Ethyl 2-oxo-4-phenylbutanoate with L-Alanyl-L-proline. Synthesis of Enalapril Maleate, J. Org. Chem., 49, 2816-2819 (1984).

6. Andeno B.V. is a subsidiary of DSM mainly active in the field of (enantiomerically pure) synthetic intermediates.

7. Sheldon, R.A., The industrial synthesis of optically active compounds. In proceedings of the chiral synthesis symposium and workshop Manchester England 18 April 1989, pp. 21-29.

8. Vriesema, B.K., ten Hoeve, W., Wijnberg, H., Kellogg, R.M., Boesten, W.H.J., Meijer, E.M. and Schoemaker, H.E., Resolution of 2-amino-5-thiomethyl pentanoic acid (homomethionine) with aminopeptidase from Pseudomonas putida or chiral phosphoric acids, Tetrahedron Lett., 26, 2045-2048 (1986).

9. Kamphuis, J. and Boesten, W.H.J., Dutch Pat. Appl. 8800260 (1988).

10. Kloosterman, M., Kamphuis, J., Schepers, C.H.M., Boesten, W.H.J., Weijnen, J.G.J., Kierkels, J.G.T., Schoemaker, H.E. and Meijer, E.M., Chiral Amino Alcohols: Two chemo-enzymatic syntheses. In Proc. 1st Int. Symp. Separation of Chiral Molecules, Paris, May 31 - June 2, 1988, Abstr. V-15.

11. Bodurow, C.C., Boyer, B.D., Brennan, J., Bunnel, C.A., Burks, J.E., Carr, M.A., Doecke, C.W., Eckrich, T.M., Fisher, J.W., Gardner, J.P., Graves, B.J., Hines, P., Hoying, R.C., Jackson, B.G., Kinnick, M.D., Kochert, C.D., Lewis, J.S., Luke, W.D., Moore, L.L., Morin, Jr., J.M., Nist, R.L., Prather, D.E., Sparks, D.L., and Vladuchick, W.C., Enantioselective synthesis of Loracarbef (LY 163892/kT 3777), Tetrahedron Lett., 30, 2321-2324 (1989).

12. Wu, S.-Y., Hirashima, A., Kuwano, E., Eto M., Synthesis of Optically Active 1,3,2-Oxazaphospolidine-2-Sulfides and 1,3,2-Benzodioxaphosphorin 2-Sulfides, Agric. Biol. Chem., 51, 537-547 (1987).

13. Wu, S.-Y., Takeya, R., Eto, M. and Tomizawa, C., Insecticidal Activity of Optically Active 1,3,2-Oxazaphospholidine 2-Sulfides and 1,3,2-Benzodioxaphosphorin 2-Sulfides, J. Pesticide Sci., 12, 221-227 (1987).

14. Eto, M., Hirashima, A., Tawata, S. and Oshima, K., Structure-Insecticidal Activity Relationship of Five-Membered Cyclic Phosphoramidates Derived from Amino Acids, J. Chem. Soc. Jpn., 5, 705-711 (1981) (Jap.).

15. Roth, H.J. and Kleeman, A., Arzneistoffsynthese, Thieme Verlag, Stuttgart, 1982, pp 376-377.

16. Wilkinson, R.G., Shepherd, R.G., Thomas, J.P. and Banghn, C., Stereospecificity in a new type of synthetic antituberculous agent. J. Am. Chem. Soc., 83, 2212-2213 (1961).

17. Francalanci, F., Cesti, P., Cabri, W., Bianchi, D., Martinengo, T. and Eoà, M., Lipase-Catalyzed Resolution of Chiral 2-Amino-1-Alcohols, J. Org. Chem., 52, 5079-5082 (1987).

18. Kleemann, A., Roth, H.J., Arzneistoffgewinnung, Thieme Verlag, Stuttgart, 1983, p. 50.

19. Takahashi, S., Microbial production of D-p-hydroxyphenylglycine, Prog. Ind. Microbiol., 24, 269-279 (1986). For classical resolution approaches, see ref. 8-20 herein.

20. Weigert, W.M., Offermans, H. and Scherberich, P., D-Penicillamin. Herstellung und Eigenschaften, Angew. Chemie, 87, 372-378 (1975).

21. See for example: Yokozeki, K. and Kubota, K., Mechanism of asymmetric production of D-amino acids from the corresponding hydantoins by Pseudomonas sp., Agric. Biol. Chem., 51, 721-728 (1987). Yamada, S., Maeshima, H., Wada, M. and Chibata, I., Production of D-alanine by Corynebacterium fascians, Appl. Microbiol., 25, 636-640 (1973). Toray Ind. Inc. Patent WO 8806.188 (1988). Asano, Y., Nakazawa, A., Kato, Y. and Kondo, K., Isolierung einer D-stereospezifischen Aminopeptidase und ihre Anwendung als Katalysator in der Organische Synthese, Angew. Chem., 101, 511-512 (1989).

22. DSM/Stamicarbon, The Netherlands and NOVO/Nordisk A.S. Denmark Eur. Pat. Appl. 0307023 (1989).

23. Boesten, W.H.J. and Peters, P.J.H., Eur. Pat. 150854 (1984). Kamphuis, J., Schepers, C.H.M., Boesten, W.H.J., Roberts, M.J.A., van Balken, J.A.M., Meijer, E.M. and Schoemaker, H.E., 'IUPAC Congress on National products', Den Haag, 1986, PB 11.

24. Taylor, S.C., Fluorinated fine chemicals by Bio-Organic technology, Speciality Chemicals, pp. 236-244 (1988).

S-2-CHLOROPROPANOIC ACID BY BIOTRANSFORMATION

DR STEPHEN C TAYLOR
ICI Biological Products
PO Box 1, BILLINGHAM, Cleveland, TS23 1LB

ABSTRACT

S-2-Chloropropanoic acid is a key chiral synthon required for the synthesis of a range of important herbicides as their single active isomers. This paper describes work done by ICI to develop a new, cost effective route to this molecule by biotransformation with a novel dehalogenase enzyme.

INTRODUCTION

The agrochemical industry worldwide is moving towards the development of more active and more selective pesticides which have minimal effect on the environment. The increasing molecular complexity which is associated with this trend is paralleled by an increasing level of attention being paid by the industry to chirality and in particular to the resolution or synthesis of single enantiomers where pesticidal activity is found to be specifically associated with one isomer form. In many cases, such as with the pyrethroid insecticides, physical/chemical methodologies have proved to be highly successful in allowing for the cost effective manufacture of specific isomers. However this technology is not always cost effective and with some product examples there is a clear synthetic role for biotechnology and more specifically for biotransformation technology.

Herbicides which are based on a propanoic acid nucleus are widely used in agriculture and S-2-chloropropanoic acid (S-CPA) represents a key intermediate for the synthesis of single isomer products. This paper will examine the background to new ICI biotransformation technology based on the application of a novel and highly stereospecific dehalogenase enzyme to the cost effective production of S-CPA.

PROPANOIC ACID BASED HERBICIDES

Derivatives of propanoic acid, such as chloropropanoic acid and lactic acid, have been, and remain, important building blocks for a wide range of herbicides, examples of which have been in agricultural use since the 1950's. The simplest of these herbicides structurally is dalapon, 2,2-dichloropropanoic acid, which has been used for the control of annual and perennial grasses. A characteristic of most of these products however is the presence of a chiral centre which arises from carbon-2 of the propanoic acid moiety. There are many examples of these products made by all of the major agrochemical companies. Dichlorprop, fenoprop and mecoprop have been in use for some time whilst fluazifop, flamprop, fenoxaprop, benzoylprop and

napropamide are more recent introductions.   Products such as fluazifop and
fenoxaprop are particularly valuable giving very effective post emergent
control of annual and perennial grass weeds in a wide range of broadleaved
crops at low dose rates (0.2-0.5 Kg ai/ha).   Conversely, products such as
mecoprop are used at higher dosage rates (1.5-3 Kg ai/ha) for the
post-emergent control of broadleaved weeds in cereals [1].

Addition to the propanoic acid group in these products varies from a
simple chlorinated phenyl group through to quite complex heteroaromatic
systems (Figure 1).

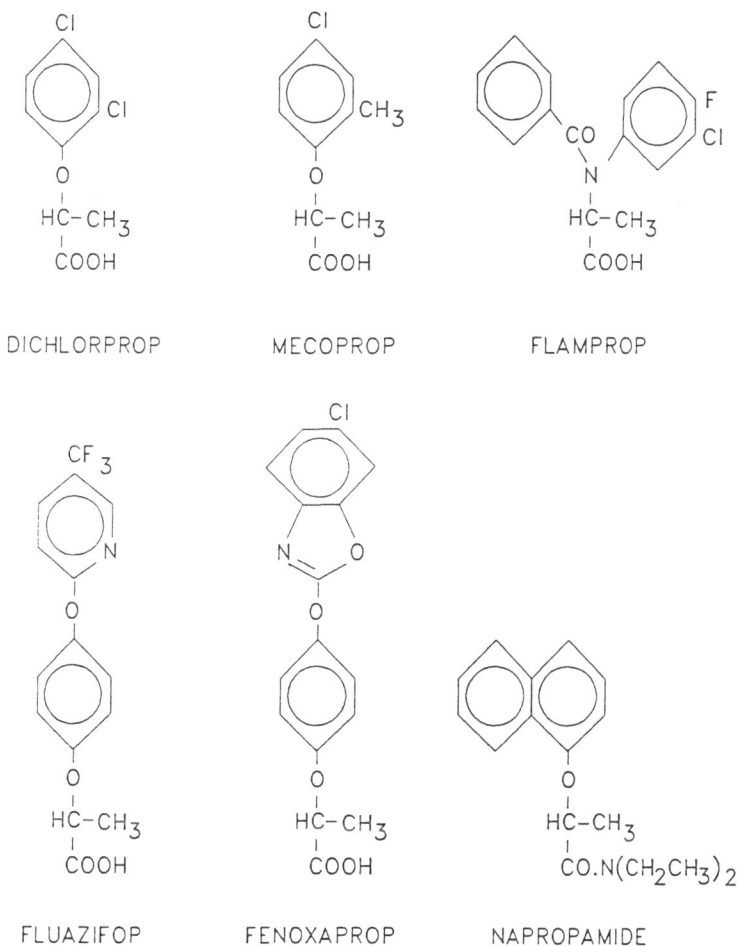

Figure 1.   Examples of propanoic acid based herbicides

The chirality of the molecule is important and in most cases the desired herbicidal activity resides only in one enantiomer. Although these herbicides have traditionally been produced as their racemates there is now a strong move in the industry to produce products as their single active isomer forms. This is happening both for reasons of reduced synthesis costs, particularly important with the complex heteroaromatic molecules, and to reduce the load on the environment of these widely used compounds.

With most of these products the chiral centre is derived from 2-chloropropanoic acid (CPA). To make the active isomer form alone S-CPA is the preferred starting material which thus represents an important intermediate for herbicide synthesis.

ENZYMIC DEHALOGENATION

Most halogenated compounds entering the environment are degraded relatively rapidly, largely as a result of microbial activity. Dehalogenase enzymes play a key role in this biodegradation and are found quite widely in nature both in prokaryotic and eukaryotic organisms often allowing their hosts to utilise halogenated molecules as sole sources of carbon and energy for growth. The largest group of dehalogenase enzymes studied in detail, are those which catalyse the hydrolysis of 2-halo substituted alkanoic acids to yield the corresponding hydroxy or oxo-acids. For example the hydrolysis of 2,2-dichloropropanoate to pyruvate, 2-chloropropanoate to lactate and chloroacetate to glycollate. These enzymes have been isolated from micro-organisms as diverse as pseudomonas, moraxella, and rhizobium all of which readily metabolise the products of dehalogenation and thus have been isolated by simple elective culture with halo-acids as carbon sources.

The halo alkanoic acid dehalogenases themselves are, in many respects, unremarkable enzymes. They require no co-factors or metal ions for activity, have temperature optima around 30° and alkaline pH optima between 8 and 10.5. Where examined, substrate specificity has been reported to be restricted to short alkyl chain chloro or bromo acids. However, dehalogenases do on occasion display one potentially valuable property, that of stereospecificity. Where chiral halo acids such as CPA have been studied as dehalogenase substrates, two specificities have been found as standard. Many enzymes show no preference for R or S enantiomers. However, at least three dehalogenases have been reported as having significant activity only with the S-enantiomer of CPA, no activity being found with R-CPA (Table 1). Prior to our work in ICI, and a more recent paper [2], no dehalogenase had been reported to be specific to R-CPA. Almost without exception, the hydrolysis reaction involves an inversion of configuration. There is however at least one report of a dehalogenase enzyme whose mechanism involves retention of configuration [3].

TABLE 1

The reported specificities of dehalogenases towards CPA isomers

Microbial	Activity		
Source	R-CPA	S-CPA	Reference
Pseudomonas sp	−	+	5
P. putida 113	+	+	6
P. putida PP3			
dehalogenase 1	+	+	3
dehalogenase 2	+	+	3
P. putida sp 109	−	+	7
Moraxella sp	+	+	8
P. cepacia MBA4	−	+	4
Rhizobium sp			
dehalogenase 1	−	+	2
dehalogenase 2	+	+	2
dehalogenase 3	+	−	2

The physiology of the dehalogenase containing micro-organisms is of great interest and as indicated in Table 1, many organisms contain multiple dehalogenases often with very different activities, specificities and even reaction mechanisms [2,3]. The significance and role of these multiple enzymes, whose combination varies with growth conditions [4], has yet to be resolved.

DEHALOGENASE TECHNOLOGY

Several biotransformation approaches to making S-CPA have been proposed and the 'standard' enzymic approach of racemic ester resolution has met with some success [9, 10]. However, resolution of CPA with most esterase and lipase enzymes does suffer from a lack of total isomer specificity shown by the enzyme which leads to reduced yield and relatively low productivity. We postulated that a racemic CPA resolution route based on an isomer specific dehalogenase, catalysing a reaction directly at the chiral centre, might offer some significant advantages in this context. Reaction of an R-CPA specific dehalogenase with CPA should lead to selective hydrolysis of R-CPA giving lactic acid and unreacted S-CPA could be isolated by one of several possible technologies (Figure 2).

In the preceding section the stereospecificity of known dehalogenase enzymes has been indicated and it has also been stated that at the time of these ideas in ICI no R-CPA specific dehalogenase had been reported. We therefore instigated a search for the missing R-CPA specific dehalogenase which would not only have the desired stereospecificity but also have those other characteristics that are essential if a cost effective, scalable biotransformation technology is to become a commercial reality.

racemic CPA

(R)—CPA
DEHALOGENASE

(S)—CPA + LACTIC ACID

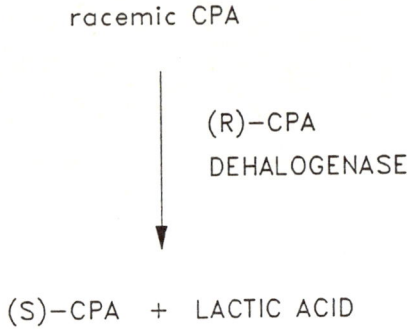

Figure 2. Rationale for the use of a dehalogenase enzyme to produce S-CPA

By the application of defined selection and isolation criteria, numerous micro-organisms of differing genera were found which contained dehalogenase enzymes. Of these, at least five contained R-CPA specific enzymes. Without exception these were all present in combination with another dehalogenase of differing substrate specificity (Table 2). Although the presence of multiple dehalogenases was not surprising it does perhaps indicate why the expected occurrence of R-CPA specific enzymes in those wild type organisms found which could only grow on R-CPA, and not S-CPA, was not borne out in practice. All five strains utilised both isomers of CPA as growth substrates.

TABLE 2

CPA isomer specificity of dehalogenases in microbial isolates.

Micro-organism Isolate	Dehalogenase I	Dehalogenase II
P. putida NCIMB 12018	R-CPA	S-CPA
P. fluorescens NCIMB 12159	R-CPA	S-CPA
NCIMB 12158	R-CPA	S-CPA
NCIMB 12160	R-CPA	R/S-CPA
NCIMB 12161	R-CPA	R/S-CPA

P. putida NCIMB 12018 was further studied.   This strain contains a low
molecular weight S-CPA specific dehalogenase and a higher molecular weight
R-CPA specific dehalogenase.   These intracellular enzymes were readily
separable either by ion exchange chromatography or on the basis of their
differing molecular weights.   The hypothesis that the R-CPA dehalogenase
would give access to high quality S-CPA, was tested in a simple temperature
and pH controlled batch bio-reactor with racemic CPA.   As expected in the
presence of a crude cell free preparation of R-CPA dehalogenase, R-CPA was
rapidly hydrolysed with no loss of S-CPA (Figure 3).   When uptake of alkali
ceased the reaction mixture was acidified, filtered and the remaining CPA
solvent extracted.   This was found to be S-CPA with an enantiomeric excess
of >98% in a yield of 48%.   An equimolar amount of S-lactic acid was formed,
also in high optical purity.

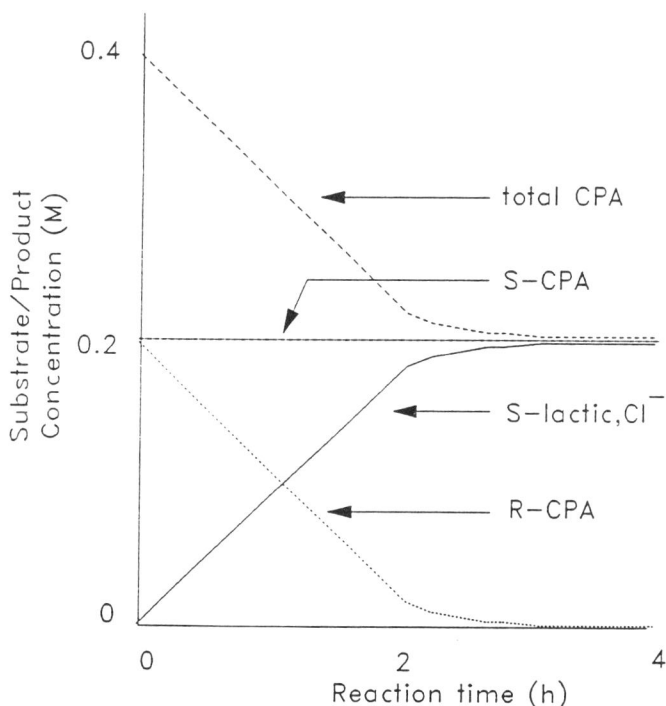

Figure 3.   Time course of a reaction for enzymic dehalogenation of racemic
            CPA using an R-CPA specific dehalogenase.   Initial concentration
            of CPA was 0.4M.   Temperature was maintained at 30° and pH at
            7.8 by titration with sodium hydroxide liquor.

Translating this encouraging result into a viable large scale biotransformation has required a major process development programme with many possible process options needing consideration. For example, the use of an isolated R-CPA dehalogenase enzyme as catalyst versus the use of mutated or cloned whole micro-organisms, free or immobilised. These different approaches had to be considered in terms of development timescales, likelihood of meeting technical targets and R&D, operating and capital cost implications.

ICI has now taken its S-CPA technology through such a multifunctional development programme and a reliable, robust and fully economic biotransformation technology for S-CPA has emerged. This developed technology is applicable to substrates other than CPA allowing some known and also some less readily available and novel optically active chiral synthons to be added to the pool for use by organic chemists in synthesis.

REFERENCES

1   The Pesticide Manual, Seventh Edition; ed Worthing CR; published by the British Crop Protection Council, 1983.

2   Leigh JA, Skinner AJ and Cooper RA (1988)   FEMS Microbiology Letters, 49, 353-356.

3   Weightman AJ, Weightman AL and Slater JH (1982)   Journal of General Microbiology, 128, 1755-1762.

4   Tsang JSH, Sallis PJ, Bull AT and Hardman DJ (1988)   Archives of Microbiology, 150, 441-446.

5   Little M and Williams PA (1971)   European Journal of Biochemistry, 21, 99-109.

6   Motosugi K, Esaki N and Soda K (1982)   Journal of Bacteriology, 150, 522-527.

7   Motosugi K, Esaki N and Soda K (1982)   Agricultural and Biological Chemistry, 46, 837-838.

8   Kawasaki H, Tone N and Tonomura K (1981)   Agricultural and Biological Chemistry, 45, 35-42.

9   US Patent 4613690 (1986) assigned to Stauffer Chemical Company.

10   Cambou B and Klibanov AM (1984)   Applied Biochemistry and Biotechnology, 9, 255-260.

BIOTECHNOLOGY IN THE FLAVOUR AND FRAGRANCE INDUSTRY

P.S.J. CHEETHAM

Unilever Research, Colworth Laboratory, Colworth House, Sharnbrook, Bedford, MK44 1LQ, U.K.

SUMMARY

Flavours and fragrances are complex mixtures of chemicals formulated to give optimal organoleptic properties. In this paper technical aspects of flavours and fragrances are discussed, including consumer trends that are being met by scientific developments, especially in biotechnology. These developments include the elucidation of the mechanisms involved in traditional processes, thus enabling the development of large-scale optimised bioprocesses. Secondly, there is the development of novel biotechnological processes for the production of various sweet and savoury flavour materials and musks and other fragrance chemicals. Examples of some of these processes are given, and also indications of likely future developments, including the impact of genetic engineering, biocatalysis, plant genetics and cell culture, and the molecular biology of flavour and fragrance perception.

INTRODUCTION

In this paper I would to give some representative examples of the biocatalytic synthesis of flavours and fragrances, so as to illustrate trends and general principles and to show some of the ways in which innovation is taking place. Thus in biotechnological research in the Flavour and Fragrance industry we ask two important questions. Firstly, 'What makes a good flavour or fragrance?' and secondly, 'How can we achieve these targets using biotechnological methods?' Thus there are two important contributions present: of market pull, i.e. what customers require; and also technology push, i.e. the new technical capabilities presented by recent advances in the biological sciences.

Socio-economic trends taking place now include demographic changes, e.g. towards older populations and changes in consumer perceptions such as the trend towards healthy living and eating that has generated a need for natural food flavours. Such market trends represent considerable challenges to the Flavour and Fragrance industry and are being met by innovative technical developments which include enzyme and fermentation technologies, and in certain cases genetic engineering and plant cell tissue culture.

The Flavour and Fragrance Industry

Before discussing the technical aspects, I first need to give some details about the Flavour and Fragrance industry in order to illustrate the role of biotechnology. The world market for flavours and fragrances has a current turnover of ca. $6B p.a. with about a 6% p.a. growth rate. Food flavours, for instance, represent about 10-15% by weight and 25% by value of the world food additive market and the world market for cosmetics is ca. $21-22B p.a. The industry is very definitely international in nature. The most important single market is obviously the USA, (ca.30% of world demand), with Western Europe contributing 35% and Japan 10%, but is very fragmented in terms of the geographic distribution and the market shares of the companies involved.

The Flavour and Fragrance industry is characterised by the use of a very large number of raw materials to produce a wide range of products, each of which is a very complex mixture of different chemicals, invariably acting synergistically. Often the chemicals responsible for flavour and fragrance properties are only present at very low concentration. Also there are very wide variations in aroma and flavour properties within any one class of chemicals such as aldehydes or ketones.

The preferred composition has been arrived at on the basis of the optimal organoleptic quality of the product and also the availability and cost of the different materials; the rarer and most expensive of which will be used only occasionally and in small quantities. Therefore the industry is characterised by dealing in relatively few large tonnage chemicals, as might be expected for an industry whose products are invariably complex mixtures of individually produced chemicals, chemically complex plant extracts, etc; and that depends heavily on the highly developed skills of perfumers and flavourists for the blending of materials to produce products with optimum smell and/or taste.

The number of naturally occurring compounds known to be important for aromas is at present about 5,000 and may ultimately be as many as 10,000. The largest chemicals used include vanillin, l-carvone, l-menthol and various musks, etc. Of the 3,000 aroma chemicals produced, it has been estimated that only about 400 are made in quantities greater than 1 tpa. About 3,000 plant-derived essential oils are known, of which about 150 to 200 are commercially important. About 3-400 materials are commonly used in fragrances. Many are cheap (ca. $10kg^{-1}) commodity chemicals such as alpha-pinene which is isolated from turpentine.

The historical development of the Flavour and Fragrance industry and the role of biotechnology are closely associated. Many important products were developed empirically, but subsequently a better scientific understanding of the biochemistry and metabolism involved has been developed, for instance in the microbial synthesis of diacetyl (butter) and methyl ketones (blue cheese) flavours, which have enabled traditional processes to be scaled-up and optimised more easily.

An active current area of biotechnological research is in the synthesis of natural food flavours, partially in response to the growing consumer demand for such natural flavours. For instance some important current food related health issues include the following:-

Non-nutritive sweeteners - calorie reduction/anti-cariogenicity
Anti-oxidants - anticarcinogenic activities
Emulsifiers - lower fat use - reduction of atherosclerosis potentials
Preservatives - to protect against microbiological contamination
Salt replacers/enhancers - to lower hypertension
Gums and starches - texturing agents to lower use of fat
Flavourings - useful in weight control. In particular good quality flavours can be used to make healthy, but rather bland and unappetising foods, containing for instance high proportions of fibre, much more attractive.

It is also important to appreciate that flavour and fragrance materials often possess additional useful functional properties such as anti-oxidant or preservative activities.

Flavour applications

The advantages and disadvantages of biocatalysis are well known, such as the regio and sterio-selectivity that can be obtained in reactions, balanced however by negative features such as the poor stability of biocatalysts. Therefore applied biocatalysis can be viewed as a struggle to utilise these valuable properties while overcoming the disadvantageous features. In particular good opportunities exist for the combined use of biocatalytic and chemical steps in the same process (bio-organic synthesis).

The importance of some key character-impact chemicals in commercial flavours and fragrances is well known. For instance, gamma-decalactone is a very important component of peach, apricot and other flavours. One method of producing natural gamma-decalactone is by the enzymic hydrolysis of castor oil, followed by the microbial degradation of the ricinoleic acid contained in the castor oil to produce 4-hydroxydecanoic acid, the immediate precursor of gamma-decalactone, which is then formed by lactonisation. This involved the selection of a microorganism capable of carrying out the abbreviated β-oxidation of the C-18 ricinoleic acid into the C-10 intermediate. Another example is the Gluconobacter oxidans fermentation of glucose into the sugar 5-ketogluconic acid, which is easily recovered in high yield and pure form by precipitation as its calcium salt. The 5-KGA can then be converted into a monomethyl furanone structure, which is an important meat and savoury flavour that is normally formed by the degradation of ATP during the cooking of meat.

The related 2,5-dimethyl-4-hydroxy furanone is one of the most important flavour chemicals being widely used in fruit flavours, especially strawberry, and also some savoury flavours. A pure source of a 6-deoxy sugar is an essential raw material. De novo synthesis is possible from fructose 1,6 diphosphate and lactaldehyde using aldolase and triose-phosphate isomerase, but this synthesis is probably not high yielding and cost-effective. Commercial production concentrates on the use of naturally occurring rhamnose (6-deoxy mannose), for instance by the selective hydrolysis of rhamnose-containing glycosides such as naringin, present in citrus materials, followed by fermentation or enzyme reaction to remove impurities such as glucose, and then heating under controlled conditions with an amino acid as a source of amino groups to produce the dimethyl furanone.

A different biotechnological approach is illustrated in a process to produce flavoured yeast extracts which function as savoury flavour blocks (mixtures of several desirable flavour chemicals). In this process yeast is treated with a carefully selected combination of enzymes to degrade the yeast biopolymers into their constituent components, for instance proteins into amino acids and peptides, RNA into nucleotides, and polysaccharides into mono and oligosaccharides. A second fermentation step using lactic acid bacteria can then be performed so as to more fully develop the flavour by, for instance, fermenting the sugars into organic acids.

Fragrance applications

Biotechnology can also be applied to fragrances. One good example is the production of ω and ω-1 hydroxyhexadecanoic acids from palmitic acid by fermentation using Torulopsis bombicola. These hydroxy acids are obtained in the form of glycolipids in which the hydroxy group has been glycosylated with the disaccharide sophorose. As a result of a metabolic study of the control of the utilisation of palmitic acid and glucose, and a mutation programme to eliminate beta-oxidation and ω-1 hydroxylation, a fermentation yielding several hundred gl-1 of sophorolipid was obtained. Hydrolysis yields the free hydroxy-acids which are then cyclised to give a mixture of hexadecanolide and methylcyclopentadecanolide with excellent properties as a musk fragrance.

A similar example is a fermentation process for the production of α, ω-alkanedoic acids (dicarboxylic acids) as precursors of macrocyclic musks. Microorganisms were screened and a strain of Candida tropicalis that produced high yields of dicarboxylic acids from C10-18 alkanes was obtained. This was then mutated to give a strain that could produce ca. 120g dicarboxylic acids 1^{-1} on a 20m^3 scale. This strain is most active on C-14 alkanes, but also reacts with saturated and unsaturated fatty acids or their esters and various fats and oils. As a result the C13 tridecanedioic acid can be produced on a 150 tpa scale and the macrocyclic musk cyclopentadecanone produced from the C15 dicarboxylic acid had excellent fragrance properties.

Other important fragrance targets are the conversion of alpha or beta pinene which is easily obtained from turpentine, into l-carvone, and the conversion of sclareol obtained from clary sage oil into Ambroxide which is a good substitute for tincture of Ambergris. A selective and high yielding microbial degradation of the side-chain to produce the required diol intermediate was achieved following a screening programme, the successful microorganism, Hyphozyma roseoniger, proved to belong to a genus new to science.

Sweeteners

High intensity sweeteners are another important product area. Aspartame (aspartic acid-phenylalanine methyl ester) was first manufactured chemically; but now an enzymic process exists using the metaloproteinase thermolysin. Thermolysin is used because, unlike other proteases, it lacks esterase activity; and esterases remove the methyl group which is essential for the intensely sweet taste of Aspartame. A second high intensity sweetener is sucralose (4,1',6' trichloro trideoxy galactosucrose). In order to prevent loss of sweetness by chlorination at the most reactive C6 primary hydroxyl group C6 protected intermediates are required for its synthesis. Synthetic routes include the use of sucrose-6-acetate and tetrachloro-raffinose intermediates. These are produced, respectively; by the fermentation of glucose into glucose-6-acetate followed by a fructosyl-transferase reaction with sucrose, and then chlorination of the resulting sucrose-6-acetate to give sucralose; or by non-selective chlorination of raffinose followed by selective hydrolysis using an α-galactosidase to give sucralose.

Many other biotechnological processes are being tested including the use of alcohol dehydrogenase or oxidases to produce aldehydes such as acetaldehyde, the enzymic hydrolysis of RNA to produce nucleotide taste enhancer, the use of lipases and proteases for the accelerated ripening of cheeses, and the debittering of citrus products by the hydrolysis of naringin by naringinase or the microbial metabolism of nomilin into less bitter products. Overall a wide variety of reactions are used to produce flavours and fragrances including hydrolysis, hydroxylation, decarboxylation, β-oxidation, oxidase, deaminase and lyase reactions.

Technological Developments

A number of developments in biotechnology are taking place which are becoming sufficiently developed for application by the Flavour and Fragrance industry. These include:

Plant culture techniques
Synthetic biocatalysts (e.g. co-factor recycling)
Gas-phase reactions
Integrated product formation and selective product recovery (e.g. membrane reactors)
Multi-phase reactors

Bio-organic synthesis

Novel-modified biocatalysts

Sterio/regioselective catalysis

Plant Technologies are of considerable longer-term potential for the Flavour and Fragrance industry. So far only two products, shikkonin and berberine have been successfully commercialised using plant cell culture. Advances required include:

i) Improved plants - produced by strain development techniques such as somaclonal variation.

ii) Cost effective plant cell tissue culture methods - including development of an improved understanding of plant metabolism and regulation. A good example is the metabolic pathway whereby L-menthol is produced by mint plants from l-limonene via L-menthone. Only ca. 40% of the L-menthone produced results in L-menthol, since some L-menthol is further metabolised by acetylation and some L-menthone is converted into d-neomenthol by an alternative pathway.

iii) The use of plant cells as catalysts of bioconversions.

Advances in Fermentation Technology will include:

1) A greater range of microorganisms suitable for industrial use obtained by screening and genetic engineering.

2) A greater understanding of and ability to control microbial metabolism and physiology.

3) Development of cost-effective genetic engineering.

4) Use of advanced bioreactors/downstream processing techniques such as extractive fermentations in which products are formed and then directly removed from the bioreactor in a single operation using ultra-filtration, ion-exchange or per-evaporation techniques, etc.

Advances in Enzyme technology will require:

i) An increased range of cheap biocatalysts available in bulk quantities.

ii) The development of cost-effective co-factor technology so as to carry out synthetic reactions.

iii) An improved understanding of factors affecting enzyme properties such as active site specificity and improved methods for the structural modification of enzymes.

iv) Biocatalysts will also move into the area of Bioorganic transformations in which biocatalysts and chemical steps are used together in the same process so as to then extend the range of effective biocatalysts, for instance to carry out oxidation, hydroxylation, reduction, carbon-carbon bond formation, dehydrogenation, dehydration reactions etc., and to obtain the benefits of sterio/regiospecific control. For instance L-menthol is the only one out of eight possible isomers with the required organoleptic properties. As a result methods of purifying L-menthol have been developed by selective enzymic esterification or de-esterification, for instance by the sterioselective hydrolysis of dl-menthyl esters by microbial cells.

In addition biological recognition effects are very important; in particular the structure-function relationships of flavour and fragrance materials are of very great importance. Sensory understanding includes receptor site mechanisms such as odour, malodour counteraction, taste, synergy effects and the structure-activity relationships of flavour and fragrance compounds, which may eventually lead to the systematic design of new aroma chemicals. Thus we have rough rules for musks and sweeteners but a more detailed and general understanding still eludes us. However, important advances are being made in the molecular biology of flavour and aroma perception such as the identification of proteins that bind the flavour and fragrance molecules and transport them to receptors.

Conclusions

In conclusion, I would like to emphasise that the Flavour and Fragrance industry is highly science-based and that biotechnology has an important role. A vital feature of Bioscience research is that it is multi-disciplinary with biochemists and microbiologists interacting with chemists, engineers, food scientists and computer scientists, etc., and that biotechnology can have important effects at various stages of a process, e.g. substrate modification, analysis techniques, product delivery, etc. We also endeavour to make our research market-led, involving marketing, buying and applications experts, etc.

References

For those readers interested in learning more, the following is a short-list of review articles dealing with the subject of biotechnology and flavours and fragrances.

Armstrong, D.W. and Yamazaki, H. TIBTECH Oct. (1986) 165.

Cheetham, P.S.J. and Lecchini, S.M.A. Food Technol. Int. (1988) 257.

Cheetham, P.S.J. in 'Biotechnology - the Science and the Business', Harwood Academic, (in press).

Gatfield, I.L. in Advs. Food Biotech. 2, (1988) 59.

Sharpell, F.H. in Comp. Biotechnol. 3, (1988) 965.

Trivedi, N. Biotechnol. Food Proc. (1986) 115.

ENANTIOSELECTIVE ENZYMATIC ESTERIFICATION OF CHIRAL ALCOHOLS USING VINYL ACETATE: A CONCEPT FOR REACHING THE DESIRED EXTENT OF CONVERSION

GERD FÜLLING, E. WOLFGANG HOLLA, REINHOLD KELLER
Hauptlaboratorium der Hoechst AG
6230 Frankfurt/M. 80, FRG

ABSTRACT

The optical resolution of a wide range of racemic secondary alcohols has been accomplished via enantioselective esterification employing lipase P as biocatalyst under nearly anhydrous conditions. The scope of structure variation with respect to enantioselection was investigated. Especially 1-arylalkanols were resolved quite effectively. Vinyl acetate was used as acyl donor. Hence, the enzyme-mediated acylation became irreversible and mostly both the remaining substrate (alcohol) and the product (acetate) but at least one fraction could be obtained in high optical purity.

INTRODUCTION

Lipases are valuable chiral catalysts for the resolution of racemic alcohols and carboxylic acids via enantioselective hydrolysis, esterification or transesterification [1]. Reactions carried out in aqueous media are shifted toward completion due to the high concentration of the nucleophile water. In contrast, in lipase-catalyzed esterifications and transesterifications carried out in organic solvents mostly neither the acyl donor carboxylic acid or -ester nor the nucleophile alcohol is present in large excess. Hence, the reaction is terminated, controlled by equilibrium, at a certain stage of conversion. Further, according to Sih et al. the kinetic treatment for the prediction of the enantiomeric excess (ee) versus conversion in kinetic resolutions becomes more

complex if the reverse reaction has to be considered [3]. The
enantiomeric excess of both substrate and product will decrease
when a certain extent of conversion is exceeded [3].

However, the application of vinyl esters as acyl donor for
the enzymatic esterification of alcohols offers an elegant
solution to overcome these problems. The enol, which is
released when the lipase-mediated attack of the acyl donor by
the substrate alcohol occurs, immediately tautomerizes to
acetaldehyde and the reaction obviously becomes irreversible
[4,5,7]. Hence, the kinetics can be treated as described by Sih
et al. for an irreversible process, e.g. the enzymatic
hydrolysis [2]. As a consequence, the optical purity of either
the substrate or the product can be enhanced by variation of
the conversion, especially if the enzyme exhibits only a
moderate enantiomeric ratio E (potential of the enzyme to
discriminate between the two enantiomers).

Based on this vinyl ester strategy, we investigated the
stereoselective acylation of several racemic secondary alcohols
employing lipase P in order to elucidate the scope of structure
variation with regard to the enantioselectivity.

MATERIALS AND METHODS

The racemic alcohols were purchased from Aldrich-Chemie GmbH &
Co KG or were prepared from their corresponding ketones ($NaBH_4$,
MeOH, 0°C). Vinyl acetate used in this investigation was a
product of Hoechst AG but is also readily available. Lipase P
(FP) was obtained from Amano Pharmaceutical Ltd and was used
without further purification.

In a typical experiment, 1-phenylethanol (rac-**1a**) (10 g)
was dissolved in vinyl acetate (100 ml) which serves both as
solvent and as acyl donor. After addition of Lipase P (1 g)
the reaction mixture was stirred at room temperature until 50%
conversion were achieved (24 h). The enzyme was removed by
filtration and was reusable. Excess vinyl acetate was
evaporated. After separation of the mixture by flash
chromatography on SiO_2 (hexane/ethyl acetate, 20 : 1), 6.4 g of
acetate (R)-(+)-**1b** (48% yield, ee $\geq$ 95%) and 4.5 g of remaining

alcohol (S)-(-)-1a (45% yield, ee ≥ 95%) were isolated.

Further reactions have been carried out on a 1 - 100 g scale. The conversion was measured by ^{1}H NMR spectroscopy. The optical purity of the products was determined either via optical rotation or via ^{1}H NMR spectroscopy using the chiral shift reagent Eu(hfc)$_3$.

RESULTS AND DISCUSSION

The kinetic resolution of several 1-arylalkanols employing different lipases either via (trans)esterification or via hydrolysis of their corresponding acetates has frequently been reported in the last few years [5-11]. Lipase P often was the enzyme of choice [5,9,10]. However, a systematic investigation to elucidate the flexibility of the enzyme toward structure variation of the substrate has not been reported yet. Therefore, we studied several racemic alcohols with extended structural requirements.

Secondary Alcohols Containing Aromatic Substituents
Substrates bearing aryl-substituents have been investigated first (figure 1). A typical reaction is presented in scheme 1. The results, summarized in table 1, clearly confirm a high potential of lipase P to discriminate between the two optical isomers of 1-arylalkanols. 1-Phenylethanol (rac-1a) was the most suitable substrate. According to scheme 1, the R-enantiomer (R)-1a but not the S-enantiomer (S)-1a was rapidly and completely converted into the corresponding acetate (R)-1b.

1-Phenylethanol derivatives **2a**, **5a** and **6a**, bearing substituents in the m- or p-position of the aromatic ring such as alkyl, alkoxy or even the electron withdrawing nitro-group did not seriously affect the highly enantioselective course of this resolution process though in the case of **5a** a decreased velocity was observed. However, a second substituent in the o-position (**3a**, **4a**) may be critical. While lipase P still was able to discriminate between the two enantiomers of rac-**4a** (E ≥ 100), the enantiomeric ratio (E = 5) drastically decreased for the enzyme-catalyzed acylation of rac-**3a**.

9a

15a

10a

16a

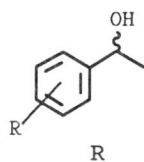

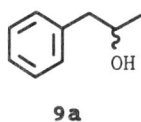

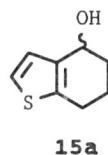

	R
1a	H
2a	4-^{i}Bu
3a	2-Me
4a	2-Cl
5a	4-NO$_2$

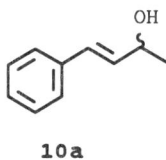

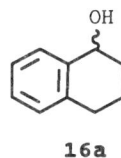

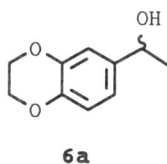

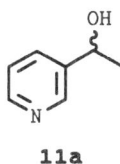

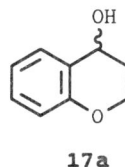

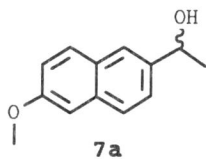

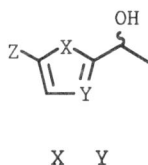

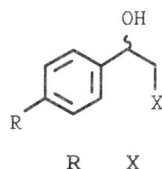

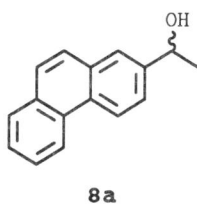

6a

11a

17a

7a

	X	Y	Z
12a	S	CH	Cl
13a	O	CH	H
14a	S	N	H

	R	X
18a	H	CH$_3$
19a	H	C$_{11}$H$_{23}$
20a	H	Cl
21a	NO$_2$	Br
22a	H	CH$_2$Cl

8a

Figure 1. Chiral secondary alcohols bearing aryl-substituents, investigated in the lipase-P-catalyzed acylation with vinyl acetate as acyl donor.

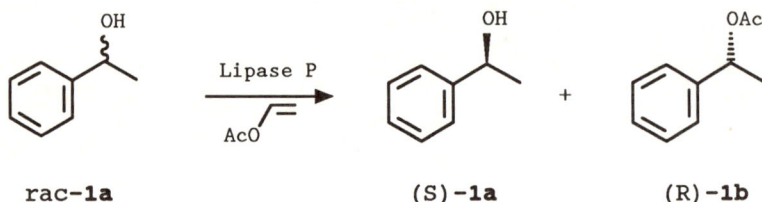

SCHEME 1

TABLE 1
Lipase-P-catalyzed enantioselective acylation of racemic
alcohols bearing aryl-substituents with vinyl acetate

alcohol	$c^{a)}$ (%)	$t^{b)}$ (h)	alcohol	ee (%)	acetate	ee (%)	$E^{c)}$
rac-**1a**	50	24	(S)-(−)-**1a**	≥95	(R)-(+)-**1b**	≥95	≥100
rac-**2a**	50	30	(−)-**2a**	≥95	(+)-**2b**	≥95	≥100
rac-**3a**	37	165	(−)-**3a**	33	(+)-**3b**	57	5
rac-**4a**	49	168$^{d)}$	(−)-**4a**	91	(+)-**4b**	≥95	≥100
rac-**5a**	50	48$^{e)}$	(−)-**5a**	≥95	(+)-**5b**	≥95	≥100
rac-**6a**	47	18	(−)-**6a**	85	(+)-**6b**	≥95	≥100
rac-**7a**	50	96	(−)-**7a**	≥95	(+)-**7b**	≥95	≥100
rac-**8a**	44	44	(−)-**8a**	75	(+)-**8b**	≥95	≥ 88
rac-**9a**	47	96	(S)-(+)-**9a**	84	(R)-(−)-**9b**	≥95	≥100
rac-**10a**	51	16	(−)-**10a**	≥95	(+)-**10b**	93	≥100
rac-**11a**	51	65	(−)-**11a**	≥95	(+)-**11b**	≥90	≥ 70
rac-**12a**	48	5$^{f)}$	(−)-**12a**	86	(+)-**12b**	92	67
rac-**13a**	52	17	(S)-(−)-**13a**	≥95	(R)-(+)-**13b**	≥88	≥ 58
rac-**14a**	34	48	(−)-**14a**	48	(+)-**14b**	≥95	≥ 63
rac-**15a**	58	23	(+)-**15a**	≥95	(+)-**15b**	70	≥ 20
rac-**16a**	56	76	(S)-(+)-**16a**	≥95	(R)-(+)-**16b**	76	≥ 27
rac-**17a**	53	16	(−)-**17a**	≥95	(+)-**17b**	≥86	≥ 49
rac-**18a**	37	96	(S)-(−)-**18a**	56	(R)-(+)-**18b**	≥95	≥ 69
rac-**19a**	38	>200	(−)-**19a**	54	(+)-**19b**	88	27
rac-**20a**	44	168	(R)-(−)-**20a**	75	(S)-(+)-**20b**	≥95	≥ 88
rac-**21a**	50	50$^{g)}$	(−)-**21a**	≥95	(+)-**21b**	>95	≥100
rac-**22a**	53	168$^{g)}$	(S)-(−)-**22a**	88	(R)-(+)-**22b**	77	22

a) conversion; b) time required for conversion as determined,
ratio substrate/lipase = 10; c) enantiomeric ratio calculated
according to reference [2]; d) ratio substrate/lipase = 0.25;
e) ratio substrate/lipase = 1; f) ratio substrate/lipase = 40;
g) ratio substrate/lipase = 2.

7a and **8a**, alcohols containing fused aromatic rings, were accepted by lipase P as well. Further, additional carbon bridges between the aromatic ring and the hydroxy bearing carbon, e. g. CH_2 (**9a**) or CH=CH (**10a**), did not affect the capability of lipase P to discriminate between the two enantiomers.

The acylation of 1-heteroarylethanols **11a** - **15a** was achieved with moderate to high enantiodifferentiation. Due to the irreversibility of the process at least one enantiomer, either the remaining substrate (**11a, 13a, 15a**) or the product (**12b, 14b**), was obtained in high optical purity simply by extending or reducing the extent of conversion.

Next, substrates **15a** - **22a** with alkyl side chains different from methyl have been studied. While the E-values still remained moderate to high, the increased sterical hindrance of the alkyl substituent led to a considerable decrease of enzyme activity in the case of **18a** - **22a**. For this reason higher amounts of enzyme were required to achieve reasonable reaction rates. However, when the alkyl side chain was fused to the aromatic ring (**15a** - **17a**) and hence, conformational fixed, again high rates of conversion were obtained. The 2-halo-1-arylethanols **20a** and **21a** as well as 3-chloro-1-phenylpropanol (**22a**) are considered to be valuable intermediates in the synthesis of optical pure pharmaceuticals [5,12]. The resolution of 2-chloro-1-phenylethanol (**20a**) using lipase P in vinyl acetate has also been reported by Oda et al. recently [5].

As far as data for the correlation of optical rotation and configuration have been available from a rough view of the literature [e.g. 5-13], in each case the R-enantiomer was acylated preferentially. These results indicate that lipase P is discriminating with respect to the steric bulk of the substituents. According to this observation, it was not surprising that due to the different priority of substituents, according to the CIP-rules, in the case of 2-chloro-1-phenyl-ethanol (**20a**) the S-enantiomer was acylated.

Aliphatic Secondary Alcohols

Though lipase P obviously prefers alcohols bearing aryl-substituents, we felt encouraged to extend our investigations to aliphatic chiral secondary alcohols (figure 2).

The results are listed in table 2. The degree of enantioselection for 2-octanol (**23a**) (E = 6) and 3-methyl-2-cyclohexenol (**25a**) (E = 5) was rather low. This was not unexpected because the substituents connected to the hydroxy bearing carbon are not too different in size. However, as the difference in the steric bulk of the substituents increased, again a moderate enantiomeric ratio was obtained, e.g. for 3-octenol (**24a**), 1-cyclohexylethanol (**26a**), 2-hydroxy-propion-aldehyde dimethylacetal (**27a**) and pantoyl lacton (**28a**). The latter are considered to be valuable chirons [14,15]. It is noteworthy that according to recently reported results lipase P was unable to achieve the reverse reaction (hydrolysis) of rac-O-acetylpantoyl lactone [14].

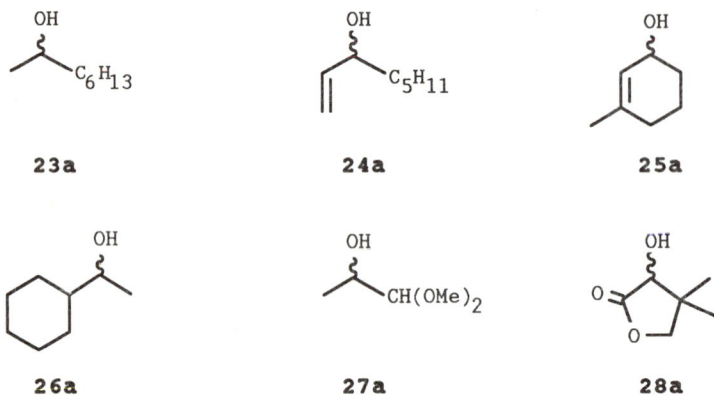

Figure 2. Enantioselective lipase-P-catalyzed acylation of aliphatic secondary alcohols with vinyl acetate.

TABLE 2
Enantioselective lipase-P-catalyzed acylation of aliphatic
secondary alcohols with vinyl acetate

alcohol	c[a) (%)	t[b) (h)	products				E[c)
			alcohol	ee (%)	acetate	ee (%)	
rac-**23a**	67	72	(S)-(+)-**23a**	85	(R)-(−)-**23b**	42	6
rac-**24a**	51	64	(R)-(−)-**24a**	86	(S)-(−)-**24b**	84	32
rac-**25a**	74	26	(S)-(−)-**25a**	88	(R)-(+)-**25b**	31	5
rac-**26a**	44	33	(+)-**26a**	74	(−)-**26b**	94	72
rac-**27a**	46	50[g)	(−)-**27a**	78	(+)-**27b**	93	65
rac-**28a**	50	336[g)	(R)-(−)-**28a**	85	(S)-(+)-**28b**	85	33

a) - g) see table 1.

CONCLUSIONS

Understanding the sope and limitations of a reaction is a major
aim in chemistry. The lipase P has proved to be a valuable
biocatalyst applicable for the resolution of a wide range of
racemic secondary alcohols. Most of these alcohols are
considered to be either valuable chirons or chiral auxiliaries.
By using vinyl acetate as the acyl donor the enzyme-mediated
acylation becomes irreversible. Consequently, the enantiomeric
excess can be enhanced by variation of the conversion. Further
improvement of the optical yields in processes characterized by
a moderate enantiomeric ratio is feasible via recycling of the
optical enriched alcohol.

REFERENCES

1. a) Jones, J.B., Sih, C.J. and Perlman, D., _Application of Biochemical Systems in Organic Chemistry_, J. Wiley & Sons, New York, 1976.
b) Carrea, G., _Trends Biotechnol._, 1984, **2**, 102-6.
c) Whitesides, G.M. and Wong, C.-H., _Angew. Chem. Int. Ed. Engl._, 1985, **24**, 617-38.
d) Kirchner, G., Scollar, M.P. and Klibanov, A.M., _J. Am. Chem. Soc._, 1985, **107**, 7072-76.
e) Jones, J.B., _Tetrahedron_, 1986, **42**, 3351-403.
f) Dordick, J.S., _Enzyme Microb. Technol._, 1989, **11**, 194-211.

2. Chen, C.-H., Fujimoto, Y., Girdaukas, G. and Sih, C.J.,
   J. Am. Chem. Soc., 1982, **104**, 7294-9.
3. Chen, C.-H., Wu, S.-H., Girdaukas, G. and Sih, C.J., J. Am.
   Chem. Soc., 1987, **109**, 2812-17.
4. a) Sweers, H.M. and Wong, C.-H., J. Am. Chem. Soc., 1986,
   108, 6421-2.
   b) Degueil-Castaing, M., De Jeso, B., Drouillard, S. and
   Maillard, B., Tetrahedron Lett., 1987, **28**, 953-4.
   c) Wang, Y.-F., Lalonde, J.L., Momongan, M., Bergbreiter,
   D.E. and Wong, C.-H., J. Am. Chem. Soc., 1988, **110**, 7200-5.
5. Hiratake J., Inagaki, M., Nishioka, T. and Oda, J., J. Org.
   Chem., 1988, **53**, 6130-3.
6. Laumen, K., Breitgoff, D. and Schneider, M.P., J. Chem.
   Soc., Chem. Commun., **1988**, 1459-61.
7. Laumen, K. and Schneider M.P., J. Chem. Soc., Chem.
   Commun., **1988**, 598-600.
8. Bevinakatti, H.S., Banerji, A.A. and Newadkar, R.V., J.
   Org. Chem., 1989, **54**, 2453-5.
9. Boaz, N.W., Tetrahedron Lett., 1989, **30**, 2061-4.
10. Bianchi, D., Cesti, P. and Battistel, E., J. Org. Chem.,
    1988, **53**, 5531-4.
11. Drueckhammer, D.G., Barbas, C.F., Nozaki, K. and Wong,
    C.-H., J. Org. Chem., 1988, **53**, 1607-11.
12. Srebnik M., Ramachandran P.V. and Brown H.C., J. Org.
    Chem., 1988, **53**, 2916-20.
13. Newman, P., Optical Resolution Procedures for Chemical
    Compounds, Vol. 3, Optical Resolution Information Center,
    Manhattan College, Riverdale, New York, 1984.
14. Glänzer B.I., Faber K. and Griengl H., Enzyme Microb.
    Technol., 1988, **10**, 689-90.
15. Bianchi D., Cesti P. and Golini P., Tetrahedron, 1989, **45**,
    869-76.

ELECTROMICROBIAL AND RELATED APPROACHES TO ANAEROBIC BIOTRANSFORMATIONS

Eurig W. James, Neil M. Dixon, Anna M. Denholm, <u>Douglas B. Kell</u> and J. Gareth Morris

Dept of Biological Sciences, University College of Wales, ABERYSTWYTH, Dyfed SY23 3DA, U.K.

Introduction and Overview

There is now a substantial interest in the exploitation of anaerobic microorganisms for the production of fine chemicals by the biotransformation of xenobiotics (e.g. Zeikus 1980, 1983, Morris 1983, Simon & Günther 1983, Simon et al 1985a, Yamada & Shimizu 1988, Morris 1989). Especially because of their ability to maintain a low redox potential in vivo (Jacob 1970, Kjaergaard 1977), due to their possession of a great many low-potential electron carriers, work with such organisms has concentrated upon bioreductions. As recently reviewed (Wong and Drueckhammer 1985, Lovitt et al 1987, Pugh et al 1988; Japanese 1988 book), another area where these organisms have assumed prominence is in the recycling of reduced pyridine nucleotides.

Many of these studies have relied upon the fermentative metabolism of these cells to produce the necessary reducing power. However, such an approach has several disadvantages: (i) there is little control over the reactions taking place, (ii) the desired end-product must be purified from the products of fermentative metabolism, (iii) the available redox potential attainable is limited by the thermodynamics of the reactions of fermentative metabolism itself. Notwithstanding, certain redox couples, for instance H_2/H^+, CO/CO_2 and HCOOH/HCHO, possess generally acceptably low midpoint potentials (Fig 1), and create or consume gaseous or otherwise non-toxic and easily-removed products, and recent attention in this area has indeed concentrated on these (White et al 1987).

A more convenient approach would be simply to generate the necessary reducing power at an electrode. Whilst several purified proteins have been shown, under appropriate conditions, to be more-or-less reveribly electroactive (Armstrong et al 1986, Frew & Hill 1988), and may usefully be exploited for biotransformations (Hill et al 1985), intact cells are for practical faradaic purposes electroinactive. Thus to effect the transfer of reducing power from an electrode to the interior of a cell, it is necessary to add one or more redox mediators of relatively low molecular weight (Fultz & Durst 1982). The

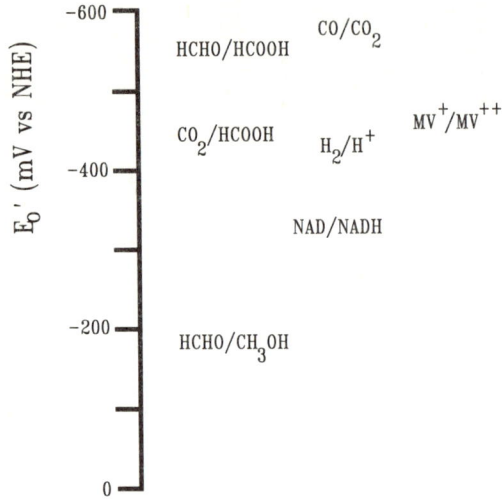

Fig 1. The redox potential of some couples of significance in anaerobic bioreductions. Data from references cited in the text and in Kell et al (1981).

properties required of such mediators, both for this purposes and indeed for use in biofuel cells (Bennetto 1984) and in amperometric biosensors (Turner et al 1987) include: (i) reversible electrochemistry at the cathode of interest, with a well-defined n value, (ii) the ability rapidly and reversibly to penetrate the cell envelope of the organism of interest, (iii) the possession of a mid-point potential suitable for the purpose intended, (iv) a lack of cellular toxicity, (v) preferably an electrical charge different from the product of the biotransformation (permitting an easy work-up by ion-exchange chromatography), (vi) easy availability at a low cost. An additional benefit of the electromicrobial approach is that the rate of the reaction of interest, when limited by the cellular reductase activity, may be measured in real time, simply as the rate of faradaic current flow.

Simon and colleagues (Simon et al 1985a,b, 1986, 1987) have shown that for chemotrophic anaerobes, the viologen dyes (Bird & Kuhn 1981) appear to be ideal redox mediators when Hg is used as the cathode; in addition, not only do viologens such as the methyl viologen (1,1'-dimethyl,-4,4'-bipyridilium) cation serve as suitable mediators at low redox potentials (<-300 mV vs SHE), but many chemotrophic anaerobes contain an exceptionally high methyl viologen-NAD reductase activity (Simon et al 1985a,b, 1986, James et al 1988). It is also worth remarking that Hg when used as a cathode, and in contrast to the Hg^{++} cation, is non-toxic to microorganisms.

Reductions of the greatest present interest include: R-COOH --> R-CHO, RCHO --> RCH_2OH and R(C=O)R'--> RCH(OH)R' (in 2 chiral forms). Further, the addition of a methylene group to a carboxylate (i.e. RCOOH + CO_2 --> RCH_2COOH), as carried out by a variety of anaerobic microorganisms, is a reductive carboxylation. In what follows, we shall give examples of each of these.

Electromicrobial techniques of use in the study and improvement of anaerobic (and indeed aerobic) biotransformations are not confined to faradaic reactions. We may here make mention of the increasingly important electroporation technique (Potter 1988), for the reversible or irreversible permeabilisation of cells for the introduction of membrane-impermeant molecules, including DNA. Finally, non-faradaic measurements ("dielectric spectroscopy") permit the real-time estimation of cellular biomass (Harris et al, 1987) and thus a rapid assessment of the toxicity of xenobiotics which may be added as substrates or solvents during biotransformations (Stoicheva et al, 1989).

Faradaic methods in anaerobic biotransformations.

As described in full elsewhere (Lovitt et al 1987, James et al 1988, Dixon et al 1989), the strategy used is to prereduce MV in an Hg pool electrode and if appropriate add it to an analytical DME, add the cells of interest, monitor any background current (due typically to hydrogenase) and then add the oxidised xenobiotic of interest, Successful electromicrobial reduction is then manifested as an increased current. A typical faradaic reaction leading to the reduction of a xenobiotic is shown in Fig 2. Here (EWJ et al, unpublished observations), cells of Clostridium La1 (Thanos & Simon 1987) were used as the biocatalyst for the reduction of the chiral beta-ketoester (Christen & Crout 1988) p-Cl-Ph-S-$COCH_2$-$COOCH_3$. The pH-dependence (Fig 3) of the reduction shows an optimum at approximately 6, where the rate of reaction (ca 200 nmol.(min.mg dw)$^{-1}$) is very respectable. The chirality of the product has not yet been determined.

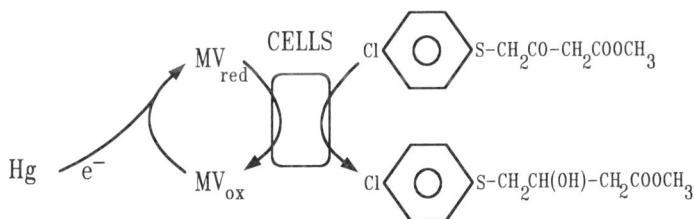

Fig 2. Reduction of a beta-ketoester using an Hg electrode, MV and cells of Clostridium La1.

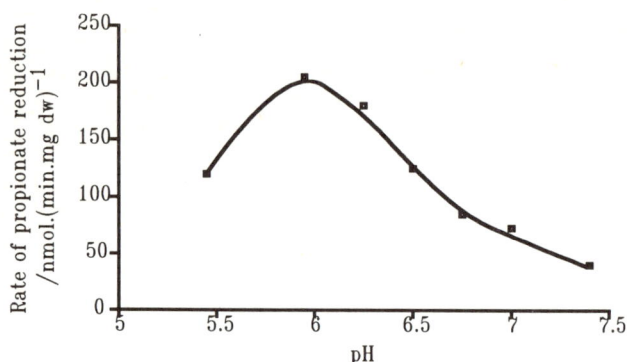

Fig 3. pH-dependence of the rate of the reaction shown in Fig 2.

We and Simon and colleagues (opera cit.) have applied this general approach (mainly with permeabilised cells) using a variety of anaerobic reductases, and suffice to say that one may ring all the expected changes with organisms (and their mode and phase of growth), substrates and environmental conditions such as pH and temperature. Rates as high as 2 umol.(min.mg dw)$^{-1}$ have been obtained for the reduction of NAD^{+} using Cl. sporogenes (James et al 1988).

Another interesting reaction is that of reductive carboxylation. For instance, Tyssee (1976) describes the purely electrochemical carboxylation at an Hg electrode of substituted acetonitriles according to the reaction RR'C(CN)H + CO_2 --> RR'C(CN)COOH. We have studied the reaction RCOOH + CO_2 --> RCH_2COOH using the electromicrobial approach (James et al 1988, Dixon et al 1989), with acetyl phosphate as the initial substrate, also in Cl. sporogenes. Because this organism contains pyruvate carboxylase, pyruvate formed may further be converted to give oxaloacetate (and, if NH_4^+ is present, even alanine and aspartate) (Dixon et al 1989). Thus it is possible to build C_4 compounds from C_2 compounds, CO_2 and electrons!

The reduction of carboxylates to alcohols is normally thermodynamically unfavourable, because of the very low mid-point potential (-550 mV) of the RCHO/RCOOH couple (Fig 1), and thus can not be driven by, say, reduced pyridine nucleotides (E_o' = -320 mV) or hydrogen (E_o' = -420 mV). However, recognising that the CO/CO_2 couple has an E_o' of some -560 mV, Simon et al (1987) and White et al (1987) incubated resting cells of Cl. thermoaceticum with CO and the carboxylate of interest, and found that when MV was present the alcohol was indeed formed with, in appropriate cases, some stereoselectivity. This reaction obviously lends itself to the electromicrobial approach, but since the CO produces the non-toxic CO_2, the approach has not

been pursued in any detail. Fraisse and Simon (1988) found
similar activities in the related Cl. formicoaceticum (albeit
with a different substrate range), and showed that formate was
another convenient electron donor. Most recently, White et al
(1989) purified the first enzyme of the "carboxylate reductase",
and found it to be a tungstoenzyme which could catalyse the
further reduction of the aldehydes formed. The great advantages
of these organisms, which are rather atypical clostridia in that
they possess electron transport chains containing menaquinone and
cytochromes (Gottwald et al 1975), is that the carboxylates are
reduced in a non-activated form and not, for instance, as CoA-
derivatives.

Although the rates of reaction and the substrate range are not
as great as one would wish, the success of the work so far
encourages one to study the system further, and we have carried
out a detailed physiological study of both clostridia in
chemostat culture (in preparation). The results of this study
lend some (limited) support to the view that the normal function
of this reaction in growing cells, as in many gratuitous
reductions of xenobiotics, is simply as an electron sink.
However, the differences between the two organisms (e.g. an
opposite dependence of carboxylate reductase activity on dilution
rate in carbon-limited chemostats) suggest that more subtle (if
teleological) reasoning should be invoked. That this type of
reaction may be far more widespread than has hitherto been
supposed is instanced by its indirect observation in
lactobacilli, which cause off-flavours of wines containing sorbic
acid (2,4-hexadienoic acid) by, inter alia, its reduction to 2,4-
hexadien-1-ol followed by a variety of isomerisations and
esterifications (Crowell & Guymon 1975).
A remarkable characteristic of anaerobic bacteria is their
ability to use as terminal electron acceptors compounds that one
does not normally even consider as redox-active (Thauer & Morris
1984, Schink 1986, Berry et al 1987). In view of the success of
reductions exploiting the CO/CO_2 couple in which the electron
donor reaction is $CO + H_2O \longrightarrow CO_2 + 2H^+ + 2e^-$, it seems
plausible that a similar reaction involving nitrogen oxides might
be expected to occur in nature, i.e. the anaerobic growth of
bacteria on NO. We do not know the mid-point potential of the
NO/NO_2 couple when participating in the analogous reaction, but
this seems an avenue well worth exploring.

Other electromicrobial techniques

In recent years, there has been much interest in the
application of relatively high electric field pulses to cells, to
induce electroporation (for the uptake of foreign, membrane-
impermeant substances) and even electrofusion (Zimmermann 1982).
We have studied ("irreversible") electroporation as a means of
rendering cells permeable to water-soluble, membrane-impermeant
substrates suitable for biotransformation, and have found it very
suitable. As an interesting sideline, we report the experiment
displayed in Fig 4, in which it may be observed that the
application (using a BTX Transfector 100 instrument) of a field
of 1 kV/cm for 3.75 ms to a suspension of Bacillus megaterium
results in an apparent increase in cell number. Since this

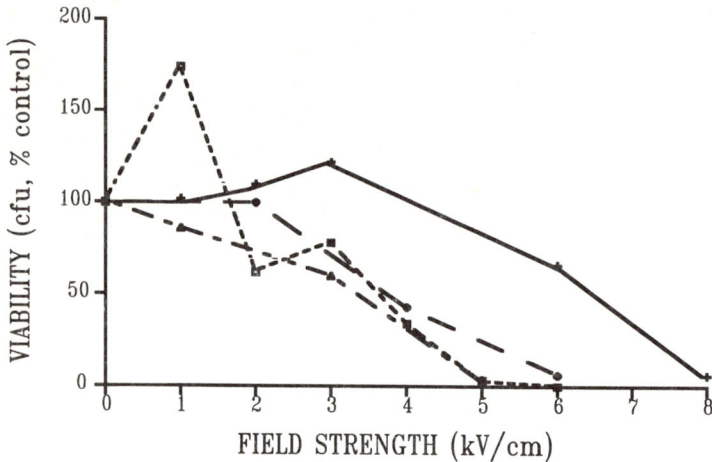

Fig 4. Irreversible electroporation (as monitored by viability) of bacteria. Cells were exposed to a 5 ms pulse of the field strength indicated in a BTX Transfector 100, and plated out on appropriate agar plates. Cells sued were <u>Escherichia</u> <u>coli</u> JMW7 (_____), <u>Bacillus</u> <u>megaterium</u> GW1 (......), <u>Clostridium</u> <u>acetobutylicum</u> NCIB 8052 (-.-.-.) and <u>Clostridium</u> <u>pasteurianum</u> ATCC 6013-MR505 (------).

species has a tendency to grow in filaments, it is assumed that the field causes **electrofission** of these filaments so as to increase the potential number of colony-forming units. Thus we may add electrofission to the possible effects induced by high electrical fields on intact microbial cells. This experiment also shows that Gram-negative bacteria are somewhat more resistant than are Gram-positive bacteria to irreversible electroporation.

Concluding remarks

In this very brief overview, we hope to have been able to show that the application of electromicrobial techniques to the study of anaerobic biotransformations has thrown up a great many most interesting features, some of which have every possibility of leading to processes of commercial interest.

Acknowledgments

We thank the Biotechnology Directorate of the SERC, U.K. and ICI Biological Products for financial support.

References

Armstrong, FA, Hill, HAO & Walton, NJ (1986) Q Rev Biophys 18, 261

Bennetto, HP (1984) Life Chem Rep 2, 363

Berry, DF, Francis, AJ & Bollag, J-M (1987) Microbiol Rev 51, 43

Bird, CL & Kuhn, AT (1981) Chem Soc Rev 10, 49

Christen, M & Crout, DHG (1988) JCS Chem Comm, 264

Crowell, EA & Guymon, JF (1975) Am J Enol Viticult 26, 97

Dixon, NM, James, EW, Lovitt RW & Kell, DB (1989) Bioelectrochem. Bioenerg. 21, 245

Fraisse, L. & Simon, H. (1988) Arch Microbiol 150, 381

Frew, JE & Hill, HAO (1988) Eur J Biochem 172, 261

Fultz ML & Durst RA (1982) Anal Chim Acta 140, 1

Gottwald, M, Andreesen, JR, LeGall, J & Ljungdahl, LJ (1975) J Bacteriol122, 325

Harris, CM, Todd, RW, Bungard, SJ, Lovitt, RW, Morris, JG & Kell, DB (1987) Enz Micr Technol 9, 181-186

Hill, HAO, Oliver, BN, Page, DJ & Hopper DJ (1985) JCS Chem Comm 1469

Jacob, H-E (1970) Meth Microbiol 2, 91

James, EW, Kell, DB, Lovitt, RW & Morris, JG (1988) Bioelectrochem Bioenerg 20, 21

Kell, DB, Doddema, HJ, Morris, JG & Vogels, GD (1981) in H. Dalton (ed) Microbial Growth on C_1 Compounds, Heyden, London, p.159.

Kjaergaard, L (1977) Adv Biochem Eng 7, 131

Lovitt, RW, James, EW, Kell, DB, Morris, JG (1987) in GW Moody & PB Baker (ed) Bioreactors and Biotransformations, Elsevier Applied Science, p. 265

Morris, JG (1983) Biochem Soc Symp 48, 147

Morris, JG (1989) in Clostridia (ed NP Minton & DJ Clarke), Plenum, London, p.193

Potter, H (1988) Anal Biochem 174, 361

Pugh, SYR, James, EW, Kell, DB & Morris, JG (1988) Biotransformations Club Report: Cofactor Recycling during Biotransformations. Laboratory of the Government Chemist, London, 35pp.

Schink, B (1986) in Biology of Anaerobic Bacteria (ed HC Dubourguier et al, Elsevier, Amsterdam, p.2

Simon, H & Günther, H (1983) in Z Yoshida & N Ise (ed) Studies in Organic Synthesis. Elsevier, New York, p. 207

Simon, H, Bader, J, Günther, H, Neumann, S & Thanos, J (1985a) Angew Chem Int Ed Engl 24, 539

Simon, H, Günther, H, Bader, J & Neumann, S (1985b) Ciba Found Symp 111, 97

Simon, H, Günther, H & Thanos, J (1986) in MP Schneider (ed) Enzymes as Catalysts in Organic Synthesis, D. Reidel, Dordrecht, p.35.

Simon, H, White, H, Lebertz, H & Thanos, J (1987) Angew Chem Int Ed Engl 26, 785

Stoicheva, N, Davey, CL, Markx, GH & Kell, DB (1989) Biocatalysis 2, 245

Thanos, ICG & Simon, H (1987) J Biotechnol 6, 13

Thauer, RK & Morris, JG (1984) Symp Soc Gen Microbiol 36, 123

Turner, APF, Karube, I & Wilson, GS (1987) (ed) Biosensors. Oxford University Press, Oxford.

Tyssee (1976) USP 3,945,896

White, H, Lebertz, H, Thanos, I & Simon, H (1987) FEMS Microbiol Lett 43, 173

White, H, Strobl, G, Feicht, R & Simon, H (1989) Eur J Biochem 184, 89

Wong, C-H & Drueckhammer, DG (1985) Bio/Technol 3, 649

Yamada, H & Shimizu, S (1988) Angew Chem INt Ed Engl 27, 622

Zeikus, JG (1980) Ann Rev Microbiol 34, 423

Zeikus, JG (1983) in Wise DL (ed) Organic Chemicals from Biomass Benjamin-Cummings, San Francisco, p 359